A ROUTE OF SURPRISES

A Soldier's Story of Strength, Service, and Discovery

LARRY R. JORDAN

Foreword by Eric K. Shinseki

UNG
UNIVERSITY of NORTH GEORGIA
THE MILITARY COLLEGE OF GEORGIA
UNIVERSITY PRESS

Published by:
University of North Georgia Press
Dahlonega, Georgia

Cover and book design by Corey Parson.
Cover photographs courtesy of Larry R. Jordan.

ISBN: 978-1-959203-19-3

For more information, please visit: http://ung.edu/university-press
Or e-mail: ungpress@ung.edu

An AUSA Book

UNG
UNIVERSITY of NORTH GEORGIA
THE MILITARY COLLEGE OF GEORGIA
UNIVERSITY PRESS

AUSA

Table of Contents

Foreword

A Route of Surprises recounts one individual's improbable journey of discovery, growth, and fulfillment during some of the most turbulent and consequential times in American history—the aftermath of the assassination of an American President, the assassination of that President's brother, the assassination of the conscience of the Nation's civil rights movement, a divisive war in Southeast Asia, the Cold War in Europe and its unfinished resolution, America's early lessons in peace enforcement, and its years long response to an attack on the American homeland on 9/11. The route he traveled instilled lessons regarding character, leadership, perseverance, and sacrifice, which led him to pursue a thirty-five year career of service to the nation as a soldier in the U.S. Army.

The author, Larry Reginald Jordan, is African American, and through his eyes, one senses what growing up in a still-segregated, post–World War II America was like and what it took to navigate the personal, professional, societal, and cultural challenges that enabled his service as a three-star, Army Lieutenant General. Throughout his journey, he was buoyed by lessons that were passed on by family, taught at West Point, and absorbed in the Army. More than an accounting of his military service alone, this story tells of discovering that he is not the first, nor the last, in his family to serve in uniform. As he began to assemble this memoir, six generations of his family had served in combat in six different conflicts, beginning with the American Civil War—a legacy of service to the Nation.

Far more than a "war story" about service in uniform, his story embraces the experiences of family—both ancestral and descendant—and friends, from private to four-star general, handyman to physician, and their influence on his journey and growth. While told through his eyes, *A Route of Surprises* is more than one man's or one family's story. It reflects the journeys of many others, who witnessed and contributed, as he did, to the remarkable emergence of the one superpower at the close of the twentieth century. I have been privileged to have known, served with, and befriended that American Soldier.

GEN Eric K. Shinseki, U.S. Army (Retired),
34th Chief of Staff, United States Army,
7th U.S. Secretary of Veterans Affairs

Prologue

It was about noon on Sunday, November 23, 1969—four days before Thanksgiving. Neither date was on my mind. I was in South Vietnam, where time took on a different dimension often measured in hours or days since the last violent encounter. It had been a few days since I'd last fired a weapon in anger.

My platoon had just reached the southwestern edge of the Michelin Rubber Plantation, having traveled from near its center. The weather was typical—hot and humid—but the monsoon rains had recently ended. As we passed the remains of a few administrative buildings, worker dormitories, and row upon row of neatly planted rubber trees, I caught a glimpse of what French colonial life must have looked like before World War II. But that imagery was far from my thoughts. Now another war, this one in Vietnam, brought all activity in the massive plantation to a halt, save that of the America-led coalition and its enemies.

I was a first lieutenant, five months into my yearlong tour, leading a mechanized reconnaissance unit of about thirty men and six armored troop carriers. That day, I was shorthanded, down to four Armored Cavalry Assault Vehicles (ACAVs). I usually had three tanks attached to provide additional firepower. Today, I would need all the firepower we could muster. A radio call from battalion command abruptly redirected us. We were to abandon our patrol of the plantation and move several miles southwest to assist an American infantry battalion in trouble. Earlier, I'd felt the deep rumble of B-52 bombers dropping half-ton bombs, strikes usually reserved for large, high priority targets. Whatever that battalion had encountered was big and dangerous.

We left the Michelin Plantation behind, and as the minutes passed, I tried to focus on what lay ahead. Faint sounds of battle began to reach us. Still, I couldn't help but wonder what awaited. We were traveling a route of surprises.

Though small, my force was formidable. Each ACAV mounted three .50-caliber heavy machine guns and carried thousands of rounds of ammunition. One vehicle was outfitted with a 106mm recoilless rifle, which was the near equivalent of a tank cannon. We could deliver the firepower of a full infantry company, though we couldn't match their manpower. Troublingly, two of our tanks had become disabled en route, and I ordered the third to remain to secure them until recovery. I couldn't wait; the order to assist had been urgent.

The lack of detail in our orders occupied my mind. The radio message gave no specifics, which was common security protocol for communication. But if nearly 600 men in a battalion couldn't handle what they were facing, what could my small force be expected to do. Other thoughts crept in. What had brought me to this moment? Was this one of those pivotal experiences that would determine the direction of my life? Or, given the circumstances, might it be the final few miles of my life?

I tried to push those unproductive questions aside. It would be many years before I understood that this road I traveled, just a few miles through the Vietnamese countryside, was part of a much longer journey. That journey included everything that had brought me to that moment and prepared me for it. I was also on a second journey—one of discovery. I would come to learn far more than I imagined about my family and the quiet, often unseen ways it had shaped me.

It would take years before I fully appreciated the similar journeys of some of my ancestors and the enduring lessons passed down through generations. Much like the road from the Michelin Plantation to the site of that battle, this other journey led into the unknown. I couldn't yet see what I would find. The outcome was uncertain, but it led me to a calling and career I now recognize as my family's business and legacy. I was traveling a route full of surprises.

CHAPTER 1

Growing Up

FEW OBSERVING ME growing up in the early 1950s in Kansas City, Missouri, would have imagined that boy as a future Army general, or as part of a multiple-generational military family. For years, I didn't know that generations before me and after me would serve in the military as part of a unique profession. Observers at the time would have envisioned a very different path through life for the young boy I was. Had they known the future, they might have wondered what led me to choose the road I eventually followed. For years, I asked myself that same question.

We were a typical African American family of the time, living in a segregated midwestern city. Though we avoided the crushing poverty common for Black Americans after World War II, our lives were neither luxurious nor even occasionally extravagant. While comfortable, money was never plentiful, and like most in my community, financial hardship was only one missed paycheck away.

My mother, Dorothy Jordan, and my grandmother, Gertrude McKim Johnson, were both schoolteachers. My maternal grandfather, Herbert Everett Johnson, had been a physician until his untimely death during the Great Depression. A 1904 graduate of Meharry Medical College, one of the few colleges then producing Black doctors, he had been the family's pride and principal breadwinner. His profession placed us within the small Black middle class, which remained financially and socially beneath nearly all Whites. After his passing, the burden of supporting the family fell on my widowed grandmother, as my mother was still in high school and years away from meeting my father. My uncle, Herbert Reginald Johnson, worked various small jobs but was unable to contribute anything financially.

My future father, Napoleon Jordan, left Oklahoma in search of work in Chicago and later in Kansas City, before serving with the Army in the Philippines during World War II. After the war, he returned to Kansas City and took a job as a postal worker following his discharge. It was there that he met and married my mother. My uncle, then a handyman and a few years older than my mother, had also served in the Army during World War II, with deployments in North Africa and Italy. Their steady and respectable employment placed our household among the more stable and admired families in the community. Still, for a Black boy like me, the possibilities and perceived future were limited by social, cultural, economic, and legal barriers that I couldn't yet comprehend.

To my young eyes, our neighborhood had everything we needed—family, friends, school, church, and even a small corner grocery store. I enjoyed my world as much as I understood it and knew it to be relatively safe and very nurturing. I was surrounded by family, friends, and adult neighbors who supported, protected, and, when necessary, corrected me. It was the kind of close-knit community where everyone knew your name, your parents, and even your grandparents. It was a community where children could play freely without fear. The people in that world left lasting impressions that I would carry with me for a lifetime. I wondered why all the people connected to institutions like school and church, but not the store, were African American. Only the store and other prosperous businesses appeared to be owned and operated by Whites. At the time I didn't question it. That was just the way things were.

None of that weighed heavily on me as a child, but what did matter were the lessons and expectations laid down at home. You were expected to do your best in school—and to do better than others. I could never hide or be inconspicuous; invariably, teachers would say things like, "I taught with your mother and grandmother," or "Dorothy Jordan would expect more from you." Perhaps they meant to inspire me. Naturally, I saw it differently and chafed under their stern looks—and the giggles of my classmates. Expectations extended beyond the home to public behavior. Neighbors often corrected or chastised the many rambunctious children doing something they'd prefer their parents not know. The common refrain in our house was, "You represent this family when you're in the street. Don't bring shame to our name!" That sense of family pride, values, and reputation took root early. Besides lectures, scoldings, or the occasional firmer reminder, we absorbed these values by listening to adults share stories from their own youth. As they sat recalling past events and relatives, we children would scamper about, only half listening—until some particularly juicy or tantalizing topic came up. When the adults chuckled wily or exchanged knowing glances and nods, our ears strained to hear or understand the veiled meaning. But if we asked questions, the answer was usually the same: "That's grown-up business." This only heightened the interest and, over time, helped shape our understanding of family history.

One such nugget of family lore involved a man my mother and uncle referred to as "Uncle Joe." When they talked about him, they'd say things like, "You know how Uncle Joe was. He was a soldier!" But whenever I asked who he was or what he had done, their answers were incomplete and evasive. Many years later, while serving at the Pentagon, I had time on weekends to dig into genealogy. The National Archives, with its census records and military documents, offered a wealth of material. Recalling those stories about Uncle Joe, I searched for military records—and the

Army, true to form, had documented everything. In the early twentieth century, recruiting officers had to telegraph the War Department on the same day they signed up new recruits. Among the records, I found a telegram dated April 27, 1905. It noted that Joseph McKim, a twenty-two-year-old, colored, with brown eyes and black hair, standing 5 feet, 5 ½ inches tall, had enlisted. He was to report to Company C, 9th Cavalry at Camp Riley, Kansas.

Seeing a photocopy of that telegram showing proof of my great-uncle's enlistment filled me with elation. Over ninety years after he joined the Army, I was a serving as a lieutenant general, feeling a deep, intimate connection to the man I had heard of as a boy. It occurred to me that Uncle Joe might have been the first in a long line of family members who had and would serve our nation in uniform: my father and uncle during World War II, myself in Vietnam and Desert Storm, and now my son, Larry Jr., recently commissioned as an armor officer and possibly headed to Iraq.

I wanted to dig deeper and found that in 1905, small unit commanders in the Army were required to submit Monthly Unit Returns. These reports detailed key events from the previous month—arrivals and departures, promotions, commendations, disciplinary actions, and more. After scanning several months of reports, I found the entry documenting Private Joseph McKim's reporting to Company C. After scanning several months of reports, I came across a note stating he had been hospitalized for minor injuries. Later, another entry reported that Private McKim was being discharged from the Army due to fighting with other soldiers. Uncle Joe had in fact "been a soldier," just as the elders had said—just not the best one. Now the conspiratorial comments and nods of my mother and uncle from years ago made sense. Nevertheless, it seemed the family's trail of military service began with him in 1905.

What I didn't know was that our family's military service began much earlier—during the Civil War. Years later, while continuing to research Great-Uncle Joe and other relatives, I discovered that Joseph McKim's father, also named Joseph, had served in Company B of the 65th U.S. Colored Infantry Regiment from 1863 to 1865. The records were fragmentary, but they revealed that after the war, he received a pension for service-related injuries, and later, his wife applied for a partial pension after his death. Learning that my great-grandfather had served in the Civil War was both a complete surprise and part of the impetus to record the family's journey more fully. Our military legacy began at a time when the nation truly needed brave defenders to step forward.

At the time I first learned about Great-Uncle Joe's past fighting, despite his small stature, I had no idea about my great-grandfather's Civil War service. That discovery would come years later. But even then, the lessons I'd learned in my childhood neighborhood remained vivid. It was a place where family stories, values, and traditions were passed down from generation to generation. Respect, family pride, and high expectations were instilled early—and reinforced often. Yet, perhaps paradoxically, so were toughness and self-confidence. While dignified behavior was expected, so was the ability to standup for yourself, even physically, if needed. I remembered one incident involving a family on our block with several boisterous children overseen by hard-drinking and ill-tempered adults. One of the boys was a notorious bully. I had to walk past that house each day after school, and after several uncomfortable encounters, I began taking an out-of-the-way route home. One day, my mother noticed and asked why. When I explained, she made it clear: either I would stand up to the bully—or I would "deal with her" when I got home.

The next day when I saw my nemesis in the front yard, I was about to turn around but I saw my mother on our front porch, with her hands on her hips and that "don't you dare" look on her face. Dorothy Jordan was slender and barely 5 feet, 6 inches tall, probably about the size of my Great-Uncle Joe. She had his same pugnacious streak, too. Rather than face her wrath, I faced my tormentor. A couple of quick jabs to the bully's face sent him off for home crying. None of those children ever bothered me again. I learned a lesson that day that remained with me: confront your fears, stand your ground, and be confident in yourself. I also learned that tough love was good love.

Not long after confronting the bully, the neighborhood and the broader world began to change. In 1954, the Supreme Court issued its landmark decision in Brown v. the Board of Education, declaring "separate but equal" unconstitutional. Desegregation soon began in Kansas City. The timing was fortunate—our old neighborhood had already begun to decline, as many stable families moved in search of better schools and a more pleasant environment. My parents decided to move, too. The prospect of new neighborhood was exciting for all of us. Still, those were tumultuous times, and emotional and social tension ran high. In previously segregated areas, Black, Puerto Rican, or Mexican families buying homes often triggered panic selling, as White residents fled. In some cases, entire neighborhood demographics shifted within just weeks or months.

In 1956, the Jordan family moved to a newly integrated neighborhood. By then, my grandmother had passed away, my uncle had moved out and married, and I had a new baby sister, Joyce. Six years my junior, I saw her as just a little kid—after all, I was already eleven and, in my mind, almost grown. Years later, Joyce would follow in our mother's and grandmother's footsteps and become a teacher, continuing what had become a family calling. Many simply assumed I would do the same. There was no sign yet that our other family legacy might be the U.S. Army.

To our surprise, the schools and residences in the new neighborhood remained integrated, with several White families staying for years. Though it was a dramatically new environment, we settled in without more than the occasional uncomfortable or less-than-cordial encounter. One incident involved two children who lived just two houses away. Joyce and I became fast friends with them—playing tag, hide-and-go-seek, and other games. But after a couple of weeks, the boy informed us that his father had forbidden them to "play with people like you." When I asked what he meant, he answered, "You know, coloreds like you." It was the first time our family had lived in such close proximity to people who weren't African American or occasionally Hispanic. This new neighborhood included a mix of races, countries of origin, and religions. Attending an integrated school for the first time, I quickly learned that skin color was not the defining factor in who was smart or dumb, nice or nasty, or athletic or clumsy. That lesson stayed with me despite the bigoted neighbor, who moved away as soon as he could sell his home.

Within a few months, I had a new set of friends and classmates. Two years later, I entered junior high school—a grand adventure filled with multiple classes, different teachers, and many new activities. The allure of "real" high school and teen years seemed tantalizingly close yet still slightly out of reach. In those rebellious years, the values I had grown up with—respect, proper behavior, hard work, and confidence—remained core parts of who I was. However, at the time, my core values weren't always visible on the surface. These lessons weren't talked about much anymore, but they had become part of us as a family. Still, among peers, no one wanted to be seen as a "Goody Two-shoes" who always followed every rule. That just wasn't cool.

In those days, I didn't think much about what I wanted to do once I was truly grown. Teaching certainly did not cross my mind. My days were spent riding my bike for hours, hanging out with the guys, and noting that girls, who had suddenly changed, were now much more interesting. Childhood dreams of becoming a cowboy, fireman, or pilot had faded. Yet whenever I watched old war movies, the life of a soldier looked heroic and exciting. I'd always liked the uniforms and had studied the various rank insignia in the encyclopedia. I read a few youth novels about the Civil War and the world wars but knew little more about the subject. I hadn't seriously thought about being a soldier—not yet.

Alongside the sweeping social changes in the country, the global order was also shifting. It was the middle of the Cold War, and I accepted the monthly air raid drills, fallout shelter signs, and duck-and-cover exercises at school as part of normal life. When the Soviets launched *Sputnik* in 1957, and later, when Francis Gary Powers was shot down over Russia in 1960 in a U-2 spy plane, the nation seemed in a panic—and even we kids sensed it. Looking back, it was a bit frightening. But for me, it was not nearly as unsettling as the separation of my parents after fifteen years of marriage. Our family became part of what would soon become a common demographic: Black, single mothers becoming head of the household, struggling to provide. My mother, who had long retired from teaching to be a homemaker, returned to the workforce as a county social worker. With her college degree, Dorothy Jordan secured a job as a welfare caseworker, dispensing Aid to Families with Dependent Children (AFDC), and other welfare assistance. But one salary wasn't enough to pay all the bills or save for college, so she also took a part-time night job. Watching her during that time, I absorbed one of life's most enduring lessons: Never give in; you can do more than you think you can.

CHAPTER 2

Finding a Direction

Central High School, with its changing demographics, prototypical teenage behavior, and location in the inner city, may seem a strange place for taking the first steps towards a military career. For its students, Central High was, in 1960s jargon, a "blast." While not always excited by the classwork, the students enjoyed their active social life and being with so many friends. Additionally, the world had greatly expanded beyond the several blocks that comprised my neighborhood. With a new driver's license in my pocket, along with a few dollars from odd jobs—and no longer needing to be in the house by the time the streetlights came on—I felt that I was finally entering adulthood. Looking around, it was easy to see the increasing maturity of my peers. Minds and bodies were growing and changing. For some, that was the problem. No longer boys and girls, young men and women grew restless, wanting to get on with life. A few dropped out of school; some got jobs; a few just hung out on street corners. Fortunately, most were doing well and had found their place in the world by the time I returned for my fiftieth high school reunion. All had matured and most had been successful in their pursuits. Still, some had not changed much. Those who had never left town for any amount of time or moved the boundaries of their world beyond where they were in the 1960s seemed to have missed something. Their world was small. However, in 1960 I had no idea just how large my world was going to become.

Thankfully, most of my friends stayed the course and graduated. For me, I had no other option than to graduate. Family tradition, values, and expectations would accept nothing less. My circle of friends were all good students who had high academic standing. We were not saints and angels. We found fun in the questionable things teens sometimes do. However, we ensured our grades were decent, as we all anticipated going on to college. It was just expected—by parents, teachers, and friends.

English, science, math, and history were interesting and easy for me. I also enjoyed drafting and mechanical drawing; I even considered becoming an engineer. Like medicine and law, engineering was a highly regarded profession—that young people were encouraged to pursue. My freshman year, I enrolled in the Junior Reserve Officers Training Program (JROTC). The program had originally been created quickly to expand the Army's officer ranks in preparation for World War I. By 1964, the version taught in high schools was academic in nature, focusing on military history, leadership, and skills like map reading, marksmanship, drill and ceremony, and service traditions. Rather than preparing students for enlistment, it was designed to instill the values of citizenship, national service, personal responsibility, and a sense of accomplishment. Students wore uniforms two days a week. Eight of the high schools in Kansas City offered JROTC, and at one time, all young men had been required to complete at least two semesters. That mandate had ended, but the tradition remained strong. Probably a quarter of the boys still took JROTC classes. I didn't realize it then, but JROTC would become a guidepost on the route I was to follow. It would help lead me to the starting point of the long journey ahead.

I found a natural fit in JROTC, with its uniforms, discipline, and growing sense of confidence it gave me. During my four years of enrollment, I earned awards and advancement. Each school with an JROTC detachment had a cadet battalion made up of three or four companies. Companies consisted of about twenty-five cadets, led by cadet captains, while battalions were led by cadet lieutenant colonels. By my senior year, I had earned the position of battalion commander and was one of the eight cadet lieutenant colonels across the city. Each spring, the high schools competed in a city-wide contest of drill and ceremonies, marksmanship, and other military skills. In addition, all the cadet battalion commanders competed for the honorary title of Cadet Colonel of the City. Although the role was ceremonial, it was considered a great honor, recognizing the top cadet in Kansas City. After intense preparation, study, a grueling test, and an interview process, I was selected as Cadet Colonel in the Spring of 1964, becoming the first African American to hold that title in the program's history. More about that later in the chapter.

The spring of 1964 brought other momentous events. The Civil Rights Movement was in full swing. President Kennedy had been assassinated the previous fall. The Cuban Missile Crisis was still fresh in people's minds. Yet for many high school seniors, the most momentous concern was college—applications, acceptance letters, and the next steps. Once again, expectations from family, friends, and others entered the picture. For me, college without a scholarship was not possible. We simply didn't have the money. Fortunately, my academic record put a scholarship within reach. I would go on to graduate fourth in the class of nearly 400 and was the highest-ranking male in what had traditionally been a female-dominated aspect of the student body at Central High.

The months before graduation were busy with college applications, essays, SAT exams, College Day sessions with recruiters, and meetings with high school counselors. After a flurry of submissions, the waiting began. Like others ranked high in the class, I received considerable advice—solicited or otherwise—from teachers, counselors, and even the principal. Everyone had a favorite or recommended college. Urged to aim high, I applied to several large Midwestern universities, including the University of Missouri and University of Kansas. I also applied to respected regional historically Black colleges and universities like Lincoln University and Central Missouri State. These schools had fine reputations and were the only options some of my friends pursued because, as they put it, they "wanted to feel at home." My last two applications were a bit uncommon, and when teachers and counselors found out, they paid attention. The advice increased exponentially. Those two schools were Yale University and the United States Military Academy at West Point.

My application to Yale stemmed in part from a family-based value: aim high, never limit yourself. The other reason was the obvious allure of the Ivy League. To students and faculty at Central High, universities like Yale and Harvard represented the pinnacle of academic achievement—more fantasy than reality.

West Point was even more distant. Most people knew little about it beyond a 1950s television series, a movie from the previous decade, or books like *The Long Gray Line*. I had read and watched them all. JROTC had only deepened the appeal—something so different, so elite, and seemingly out of reach.

The final push came from conversations: one with the school vice principal, and another with a Presbyterian minister my mother had met through her work as a welfare case worker. He had graduated from the Academy in 1945. Father Andy Carhartt saw something in the eighteen-year-old wearing a JROTC uniform and encouraged me with sincerity and persistence. Interestingly, I wouldn't learn until more than fifty years later that his nephew had been a fellow cadet at West Point, two years ahead of me. The routes we take often intersect with others in the most unexpected ways.

As April turned to May, the college acceptance letters began arriving. Most of my classmates made their decisions quickly, even when they had more than one choice. I had received a couple letters of acceptance, but only modest scholarship offers, not enough to cover costs. Then came the unforgettable day when the Jordan household received word from Yale: I had been awarded a full four-year scholarship to attend school in New Haven, Connecticut. My mother and sister were ecstatic. The Central High faculty was abuzz. My classmates' reactions ranged from congratulatory to stunned. Inside, though, I was torn. I still hadn't heard from West Point. When asked if I had accepted Yale's offer, I would reply that I was waiting to see what else might come through. Eventually, it became obvious: the only remaining possibility was an appointment to West Point. Congressman Dick Bolling had nominated me for a competitive appointment, meaning I had to be selected over the other candidates from his congressional district by the Academy's Admission Board.

As the days passed, pressure mounted. Counselors and teachers argued that Yale was an honor—not just for me, but for the entire school. Others said it was an honor for the city. Perhaps most unfairly, some argued that it was an honor for all Black people in the state, and that I had a responsibility to accept the offer.

But in her steady, maternal way, Dorothy Jordan simply said she supported whatever decision her son thought was best. Years later, I came to believe she probably favored West Point for the same reason I did—it was different, unique, and for African Americans in 1964, almost unheard of.

I wanted to wait for an acceptance letter from West Point, and one defining incident cemented that decision. Being named Cadet Colonel for Kansas City after winning the city-wide competition had given me hope, especially as the days dragged on without an acceptance letter. I was excited when the Army Reserve major serving as the Professor of Military

Science for the Kansas City School District asked to meet me downtown at the Board of Education. When I arrived, he gave me a perfunctory congratulations for being named City Cadet Colonel, something I found odd since he hadn't reached out when I first won the competition. Then he got to the point: he had heard I had applied to West Point. I confirmed that I had. His response was chilling in its bluntness, "You should reconsider. West Point isn't for you." The meeting ended five minutes after it began. I headed home confused and disappointed. Only later did I realize what he had meant: that West Point wasn't for an eighteen-year-old Black youth who was trying to complete with "deserving" White boys. I'd learned another lesson. You couldn't just be good and perform well—you had to be better than everyone else. And even that wouldn't aways be enough.

But the more I thought about it, the less I cared. The values I had learned at home made it clear I was supposed to work harder than others anyway. That officer's words didn't deter me, in fact they confirmed that I had to go. I asked Yale for an extension on the scholarship decision, and they granted it.

My appointment to West Point arrived one day before the extended deadline. Another lesson: If you believe a decision is right, don't give up!

I had found my direction—but I had no idea how far it would take me or what I would discover along the way.

CHAPTER 3

West Point

MANY FRESHMEN CONSIDER college a different world from anything they've known—but West Point was a different universe entirely. Nothing could prepare you for it. Every cadet encountered new and often bizarre experiences. For me, that was doubly true. The month between high school graduation and my July 1 reporting date at West Point, New York, flew by. So many firsts had to be planned, beginning with the journey itself. For a young man who had never traveled beyond the Midwest or more than three states from home, getting to West Point felt like an adventure. It was also my first time on an airplane. I flew into New York City the day before I was to report. The Academy had sent a letter explaining how to get from the city's airports to West Point, fifty miles to the north. It sounded simple—on paper.

In reality, it wasn't so easy for an eighteen-year-old alone in New York City for the first time. After landing at JFK Airport, I took the commuter bus to the Port Authority terminal in midtown Manhattan. Arriving in the heart of "the city that never sleeps" was almost overwhelming. The lights, noise, crowds, and energy were more than I'd ever imagined.

I was relieved to find a half-dozen other young men, equally starry-eyed, also headed to West Point. Our exuberance was soon deflated by the news that the last bus to the Academy had already departed. Alone in a huge, unfamiliar city, not knowing anyone or having a clear direction, we quickly shed our bravado. Worldly airs and cocky attitudes evaporated into silence and consternation. None of us had more than a few dollars—enough for snacks and bus fare, but not for a hotel room in New York. We resigned ourselves to spending the night in Port Authority. The metal benches were hard and cold. The public restrooms were unpleasant. The terminal was dotted with drunks, suspicious characters, and hurried travelers. Shops and food counters were shuttered. It was, for all of us, the longest, strangest, and most uncomfortable night we had ever experienced. And hanging over it all was the uncertainty of what awaited us fifty miles up the Hudson River.

Morning finally came. Two hours after leaving Port Authority, the bus delivered us to one of the most striking places we had ever seen. West Point sits on a dramatic bend of the Hudson River, atop a flat projection of land fortified during the Revolutionary War and continuously garrisoned by the Army ever since.

A broad manicured parade field called the Plain stretched between the cliffs above the river to the east and forested hills to the west. Stone barracks and historic buildings bordered the field, giving the post a timeless, even marvelous character. The view up the Hudson Valley—miles of river framed by wooded slopes—was unforgettable. We stood in awe. But within days, we would scarcely notice the view. Our attention would be consumed by the intense, alternate universe of the Academy. From that moment forward, we were no longer just young men—we were "New Cadets," about to enter an entirely different reality.

After a whirlwind of in-processing paperwork, each New Cadet received a large paper tag bearing their name and company assignment, which was hung from their belt. Then came haircuts—rapid and identical. Barbers would joke, "How do you want it?" before delivering the same ninety-second buzz cut to every head.

Next came a time-honored West Point tradition. In small groups, we were marched toward a sallyport, an arched entrance. The sallyport led into the Central Area, a large rectangular courtyard surrounded by the Central Barracks. Like many structures at the Academy, the Central Barracks (dormitories built in 1847) were made from stone in a fortress-like Gothic style. The barracks had been updated with electricity, plumbing, and steam heat, but retained the atmosphere of a nineteenth-century military institution. In the middle of the courtyard stood a twenty-foot-tall iron pillar topped with an ornate iron clock. Its four faces pointed in each cardinal direction so time could be read from all sides. While there were other barracks at West Point, none matched those around the Central Area's stark history and austere presence.

As we approached the sallyport, a roar greeted us: sounds like dozens of screaming demons mixed with excited replies, running, and other noises. We were instructed to keep our heads and eyes straight forward, say nothing, and await instructions. One by one, we were each directed to approach the sternest, most perfectly postured, and impressive cadet we had ever seen: the "Man in the Red Sash," the Cadet First Sergeant, whose crimson waist sash marked him as the official point of contact for incoming cadets. The drill was straightforward: march up, stop three paces away, salute, and say "New Cadet Jordan (or another New Cadet) reports to the man in the red sash as ordered."

I had experience with saluting and reporting to superiors from my time in JROTC, but nothing prepared me for this emotionally charged moment. Surrounded by the din of a half dozen other New Cadets shouting their reports, stumbling through instructions, and being sharply dressed down, I too faltered. By tradition, no New Cadet in history has ever dropped their bag quickly enough. No one reported properly. Each of us was loudly corrected, made to repeat the process, and finally dismissed to our assigned rooms only after meeting the Man in the Red Sash's impossible standard.

The rest of the day was a blur. We were rapidly issued uniforms—including underwear, toiletries, socks, shoes, hats, and more. Everything we had brought from our prior life was packed away into our lone piece of luggage and locked in storage. From that moment forward, everything we wore or used was government issued.

The hours passed with rudimentary training in drill and ceremony—basic marching—so the New Cadets could be marched to Trophy Point, a scenic overlook at the edge of the Plain, in a formation somewhat more organized than a mob, for our formal swearing-in as cadets and members of the U.S. military. That solemn ceremony was followed by the strangest meal any of us had ever experienced.

We were marched to Washington Hall, a huge dining hall which could seat and feed the entire West Point Corps of Cadets in under an hour. Meals were a ritual and part of the indoctrination and training for New Cadets. At each ten-person table, New Cadets sat at rigid attention, perching on the front 6 inches of the chair and staring only at their plates. We were required to eat "square meals:" raising a utensil, taking one small bite, returning the utensil to the table, placing hands back in laps, and chewing before repeating the process. Beverages were consumed in the same manner.

Upperclassmen sat at the same tables, eating normally, and interrupting the New Cadet's at any moment—correcting our posture, questioning us, or demanding we recite memorized knowledge. We had to be prepared to answer questions about everything from famous military quotes from characters like Douglas MacArthur to obscure trivia, like how many gallons were in the reservoir when full. We also had to know how many days remained until the First Class would graduate—or until Army was scheduled to beat Navy in the annual Army-Navy football game.

Every New Cadet was issued a small booklet called *Bugle Notes*, filled with quotes, acts, mottos, and traditions. Memorizing it was a requirement. This regimented eating style, complete with constant interruptions, would be our mealtime experience for the entirety of Plebe

Year. After that first dinner, we returned to our rooms—each shared with two other New Cadets—and scrambled to organize everything according to the room setup manual's detailed instructions. Every item of clothing had a designated place, and each had to be folded and displayed in a specific way. We were far from finished when the mandatory lights out order was passed at 10:00 p.m. We soon learned that 10:00 p.m. was a good thing, as reveille, accompanied by bugle and drum, came at 5:50 a.m. Thus ended the first day of New Cadet Basic Training—known in cadet parlance as "Beast Barracks."

Beast Barracks lasted two months and consisted of many hours of marching, physical training, military traditions and customs, weapons handling, and basic field instruction. Even when not training, New Cadets were always in performance mode. Outside the barracks, we had to "brace," which meant assuming an exaggerated position of attention with the chin pulled back against the neck and eyes locked straight ahead. Any deviation brought swift and severe discipline. We jogged—or "doubled timed"—everywhere. No walking allowed.

Beast Barracks often has been likened to pledging a fraternity for sixty straight days, twenty-four hours a day. But the reward was tangible: if you made it through, you were officially accepted into the Corps of Cadets—not just a New Cadet, but a Fourth Classman, or "Plebe." Above us were Third Classmen (sophomores), known as "Yearlings," who had survived Plebe Year and took great delight in ensuring we did too—on their terms. Second Classmen (juniors) were called "Cows." First Classmen (seniors) were the highly envied "Firsties," on the cusp of graduation and commissioning as second lieutenants. Firsties had many more liberties, known as privileges: car ownership (starting second semester), taking a few weekend leaves, and even evening movie passes during the week. Some Cows served as squad leaders and held the rank of Cadet Corporal. Firsties served as sergeants and officers—the senior leaders of the Corps. This was the hierarchy we operated under.

In early September, the day finally arrived when the Class of 1968 was formally accepted into the Corps of Cadets. The trials of the past two months were reviewed in a moment of reflection. My classmates and I, had by then stopped thinking of ourselves as "Larry," "Bill," or "Joe." Now it was "Cadet Jordan," "Mister Jordan," or simply "Jordan." While we had eagerly anticipated that day, little changed in reality. Though we were no longer required to double time everywhere, the full Corps of Cadets was back on campus, which meant there were probably ten times the previous number of upperclassmen to help train and discipline Plebes. Plebes were routinely "hazed:" varied sorts of harassment applied by upperclassmen to correct minor errors or indiscretions, real or perceived. The meal routine also changed, but not for the better. Rather than eight new cadets at a table with two upperclassmen, now three Plebes sat at a table with seven upperclassmen—which meant more watching eyes and opportunities for our meals to be interrupted. During Plebe year, the Class of 1968 saw many great meals but consumed far fewer. The dining hall had remarkedly good food considering it served 2,400 cadets three meals a day, each in under an hour.

Academics had also begun, with a rigor all its own. Alongside the discipline and hazing, Plebes—like all other cadets—were expected to be ready for each academic class every day. In addition, instructors—who were active-duty officers with graduate degrees—either administered formal exams or recorded in-class oral responses to grade each cadet daily. The Class of 1968, like all others before and since, began to realize that this was to be their existence until the following June: two tough semesters, accompanied by daily classes, physical education, mandatory intramural programs, two or three weekly formal parades, and study—not to mention the omnipresent threat of hazing by either well-intended or bored upperclassmen.

Much has been written about West Point and cadet life. My own experience was similar to that of thousands who came before me—except in one respect: I was African American. That meant I could never achieve the universal goal of every Plebe: to blend in, be inconspicuous, and avoid the attention of upperclassmen. It reminded me a bit of the way I could never escape attention in elementary school.

To fully appreciate what it meant to stand out at West Point in the 1960s, one must understand the composition of the Corps of Cadets at that time. The Class of 1968 entered with 991 members—the largest entering class to date. A quirk in federal law had historically limited the size of West Point to 2,400 cadets, while both the Naval and Air Force Academies were authorized to enroll 4,000—even though the Army was the largest branch of service. Noticing the discrepancy, President Kennedy had worked to have the law changed. By the Class of 1968's graduation, the West Point Corps of Cadets had grown to nearly 4,000.

Of the 991 who entered in July 1964, only 708 graduated in June 1968. Those who had left the Academy did so for many reasons: some struggled with the intense discipline, physical rigor, academic demands, or the overall experience; others were injured and discharged for medical reasons; a small number violated the Academy's Honor Code, which specified that "a cadet will not lie, cheat, or steal, nor tolerate those who do." Those found guilty were summarily dismissed. Among the 991 who started, only ten were African American. Our arrival brought the total number of Black cadets at West Point to just nineteen. Escaping notice or attention was not possible for any of us.

A full recounting of my four years at the Academy could fill several volumes. Some memories are humorous, others sobering, and many remain surreal. Cadet life followed a strict and exhausting routine.

Every day began with a 6:00 a.m. reveille formation outside—rain, snow, or shine—followed by breakfast. Attendance at all three meals was mandatory, and we had to march to the dining hall in formation. Academic classes ran from 8:00 a.m. to 3:15 p.m., followed by mandatory intramural athletics two days a week for those not on intercollegiate sports teams. On two other afternoons, and again on Saturday after morning classes, cadets participated in formal parades. While the public delighted in watching the parades, we cadets viewed them as yet another intrusion into already extremely limited free time.

After dinner, cadets were expected to study in their rooms or in the library. Only First Classmen—seniors—had the privilege of taking weekend passes, departing post after Saturday's parade and returning Sunday evening. Late Saturday afternoon until dinner on Sunday was free time for most other cadets, except for the mandatory Protestant, Catholic, and Jewish chapel and synagogue services.

Despite the strict structure, cadet life was not devoid of mischief. Like many college students, we indulged in occasional juvenile horseplay—careful not to be caught by the officers who monitored our behavior. The atmosphere sometimes resembled the boarding schools' antics of nineteenth-century Britain. Sometimes outlandish acts provided welcome breaks from our full and disciplined days. One such act involved "borrowing" a cannon barrel from Trophy Point and under the cover of darkness secretly hauling it into an upperclassman's bed. Returning to find a 300-pound piece of artillery in their room made for quite a surprise—an uninvited bed partner that two roommates couldn't move alone.

Even the officers on the staff and faculty weren't immune. During my Plebe year, our company Tactical Officer, a Navy exchange officer, became the target of spirited ribbing during the week leading to the annual Army-Navy football game. While he was serving as overnight Officer-in-Charge, his navy-blue Volkswagen Beetle mysteriously disappeared from the parking lot. By morning, it had been carefully relocated to the lobby of Washington Hall, the cadet dining hall. A large, handmade "Beat Navy" sign was taped to the windshield.

While such moments provided entertainment, they were only brief interruptions in my demanding and transformative military journey. Looking back, I can say with certainty that West Point was the single most important, formative experience of my military career. Beyond any technical skill, it instilled the core values of military service: the ethics, discipline, and most importantly, the foundations of leadership.

West Point provided a broad and exceptional education. At that time, all cadets followed the same engineering-focused curriculum; there were no majors, and every graduate received a Bachelor of Science in General Engineering. A limited number of electives were offered, and I chose to focus on international relations and history. But some of the most valuable lessons we learned weren't academic—they were social, cultural, and emotional, especially those involving human nature. High on the list of lessons learned was the value of teamwork and camaraderie. A frequently-heard phrase was "Cooperate and Graduate." My classmates and I quickly learned what that meant: the challenges at West Point were too great to overcome alone. The team was more important—and more resilient—than the individual. We were beginning to learn the principles of selfless service and working for the greater good. Acceptance and respect within the Corps were based on how a cadet performed under pressure, maintained cool-headedness, persevered, and placed the team ahead of self. For me, most of those things were easy, being rooted in what I had learned at home.

For the ten African Americans who entered with the Class of 1968, the route to acceptance and success was much the same. Despite their disparate backgrounds, most cadets set aside superficial differences to bond with classmates who showed motivation as team players. This did not guarantee harmony and absence of personal prejudices, but by and large, Black cadets were accepted by the majority—and truly befriended by many. Still, social dynamics set us apart. That had long been true for African Americans attending West Point. But with nine Black cadets in a single class—an unprecedented number—we experienced something different. Between 1877 and 1884, only four Black cadets graduated. There were none after that until Benjamin O. Davis Jr. in 1936. In many classes after that, there were only two or three. Those smaller numbers made it difficult to meet and interact. For the first time, our cohort allowed for mutual emotional and cultural support from those who could fully relate to unique cultural backgrounds and experiences.

For each, this was an additional source of strength. Early on, we made a promise: we would not let one another become despondent, fail, or be dismissed, if within our power. Four years later nine of the ten graduated and were commissioned. The one who didn't make it through departed after the first semester. Because he didn't associate with any of us, his reasons for leaving were never clear to us. Among the nine who remained, the bond, the camaraderie, and the brotherhood between all was strong, continuing long after graduation. It was a reminder of earlier lessons about the enduring value and importance of family and friendship. In many ways, these nine men became like an extended family. I was to see that phenomenon play out throughout my military career.

These lessons—about grit and determination, honor and integrity, and knowing oneself—were the most important in propelling me into a life of service. In so many ways, they echoed the values I'd grown up with, including hard work, achievement, self-confidence, pride, and dignity. Like most West Point graduates, I truly bought into the concepts emphasized in the Academy's motto: Duty, Honor, Country. We rarely spoke seriously about it openly, but it was always present. If you didn't believe in those words, you probably wouldn't have endured the countless challenges—large and small—that we faced every month, day, and hour at "the School," as many cadets and alumni often call it.

Few symbols represent West Point more powerfully than the class ring and the yearbook, *The Howitzer*. West Point was the first institution in the United States to adopt the tradition of class rings, beginning when the Class of 1835 decided to have one made for each member of the class. For graduates, the ring is a cherished emblem—an enduring, symbol of membership in the "Long Gray Line"—the body of all who have called themselves West Pointers, past and present. Each ring features the Academy crest on one side and the class crest on the other. I had the privilege of representing my Cadet Company, B-2, on the Ring and Crest Committee. We were responsible for designing the class crest, securing class approval, and selecting the jeweler to manufacture the rings. Our design was traditional: an eagle grasping the number "68" in its talons, flanked by two cadet sabers. The day we received our rings, just months before graduation, was truly memorable. What I never envisioned then was that I would wear that ring through the jungles of Vietnam, along the Berlin Wall, across the desert in Operation Desert Storm, and within the halls of the Pentagon.

Another milestone came in the spring of 1968: branch selection. We chose our commissioning branches late in the second semester. Our summer training had introduced us to the different branches of the Army, most notably the combat arms—Infantry, Armor, Field Artillery, and Engineers. Fewer slots were available in support branches like Signal Corps, Military Intelligence, Transportation, and Quartermaster. I had become fascinated by Armor and Cavalry, greatly enjoying the hands-on training and the distinct esprit I saw among its officers and soldiers. The class ranked cadets by grade point average, and selections were made in order from the highest to the lowest. Because I was in the middle of the class, I secured my first choice. Unfortunately, those near the bottom had no choice—some were left with only Infantry. For me, it was the continuation of a journey already begun—and to my delight, it meant riding tanks.

The Howitzer chronicles the highlights of the academic year and includes photos of every cadet in the Corps. Graduating seniors receive a special recognition with large, individual portraits and short messages about each, written by one or more classmates. These notes often include jovial comments about misadventures—struggles with academics, athletic feats and accomplishments, the fame of fabulous dates, or great tastes in new automobiles for graduation. My short note contains some of those but ended with what seemed an ironic joke: "General Jordan, as indeed someday it shall please one to address him." At the time, it seemed an amusing exaggeration—but twenty-two years later, I would be addressed by that very title. Equally prophetic was the inscription I chose for my yearbook cover, embossed in gold leaf on the cover, alongside my name: "THE END AND THE BEGINNING." I wasn't sure why I selected those words at the time. In hindsight, they seemed very appropriate.

July 1, 1964–June 5, 1968: it had been three years, eleven months, twenty-five days. So much had happened. So much had changed. So much more would change. It was the end of something special. It was also the beginning of something even more special.

CHAPTER 4

Life as a Lieutenant

AFTER SIXTY SPLENDID and restful days of leave following graduation, I joined approximately fifty of my West Point classmates—fellow newly-commissioned armor and cavalry officers—at Fort Benning, Georgia, for the Ranger Course. Although the majority of our graduating class was scheduled to attend, the school could only accommodate limited numbers at a time, given the influx of trainees from across the Army to train. Therefore, newly commissioned officers were grouped by their branch for attendance.

Ranger School is one of the Army's most physically, mentally, and tactically demanding courses. Spanning eight weeks, it consists of hiking, mountain climbing, rappelling, and patrolling through forests, hills, and swamps. It was the toughest physical challenge most of us had ever faced—which was saying something, considering the rigors of West Point. There, cadets had participated in boxing, wrestling, hand-to-hand fighting, gymnastics, and survival swimming, done in the water wearing uniforms, boots, and carrying rifles. On top of the Army's standard fitness tests, West Point had its own special torture exam—the infamous Indoor Obstacle Course: a series of grueling, timed obstacles. These obstacles included climbing walls, barbed wire crawls, swinging tires, tunnels, ropes, and a final sprint around an indoor track carrying a weight. Even varsity athletes hated and sometimes struggled with the course.

Ranger School offered no such indoor challenges—it simply used Mother Nature. Mountains, swamps, and dense forests tested both endurance and persistence, while fatigue, discomfort, and hunger were constant obstacles. The course tested whether a trainee could maintain focus and mission readiness under harsh conditions. Missions typically included long-range patrols or operations such as reconnaissance or attacks on designated objectives. Often, the only way to reach the next day's food was to complete the day's mission. Students were evaluated during each exercise—sometimes as leaders, but other times as team members. Being able to perform well in both roles was important.

Ranger School was an exceptional leadership course. It wasn't just about leading missions—it was about discovering how you responded when cold, wet, hungry, and exhausted, and how others did as well. I learned how to motivate people under pressure. I learned how to persist when a task seemed impossible, and how good leaders found a way forward. In hindsight, I realized Ranger School distilled and intensified lessons I had already begun to master—lessons instilled by my family, especially my mother: never give up, strive for excellence, aim high, have self-confidence, confront your fears. At the time, my greatest satisfaction, however, came from the black and yellow Ranger tab that would be sewn on the left shoulder of the uniform, above the unit patch. That tab was recognized by all soldiers as the badge of a warrior who had successfully made it through one of the toughest courses the Army could devise.

My teammate during the Ranger Course, also known as my "Ranger Buddy," had been one of several roommates I'd had at West Point. Cadets were required to change roommates twice a year, a practice meant to encourage bonding with as many classmates as possible and to teach us how to live closely with people who might differ from us.

Steve Herman was older than most of our peers, having previously served in the Army as an enlisted soldier. He was Jewish and from New York City. Even as a Plebe, classmates jokingly called him "the old man."

At twenty-one or twenty-two years old, he was noticeably older than typical eighteen- or nineteen-year-old Plebes. He had already seen the *real* Army—and some of the world! Fortunately, we didn't encounter any difficulties beyond those that every trainee faced: grueling patrols, mountain climbing and rappelling, and water obstacles—we took all parts of the course in our stride. The greatest challenges were hunger and fatigue. To pass the time, we quietly played a guessing game on patrol: "What type of food do you miss most?" During the mountain phase, our rations were supplemented when we passed an apple tree just across the fence line of the training area. We figured the owner wouldn't miss two small, unripe apples. Another time, while digging a fighting position in sandy soil during the last phase of training, we unearthed a small radish. We carefully sliced it with a bayonet and added it to the canned combat rations that made up our meal that day.

Steve and I enjoyed a strong friendship, and that camaraderie helped us succeed. After the course, we were both headed to Fort Knox and decided to make the drive together—Steve drove his dark green Corvette, and I drove my olive-green Firebird. Both cars had New York plates. In 1968, a Jew and a Black man diving from Georgia, through Alabama to Kentucky in new sports cars were bound to draw attention. We joked about it, telling ourselves, "We're Rangers and lieutenants in this man's Army." We had both volunteered for Vietnam and would deploy after six months at our duty stations. Still, we were secretly apprehensive. While neither of us had experienced overt discrimination at West Point or in the Army, we weren't so sure that would hold true across the rural South. Somewhere in rural Alabama, we had to stop for gas to satisfy the two large V8 engines powering our dream vehicles. While gassing up, we saw the service station proprietor coming outside. One glance told me all I needed to know: bib overalls, red handkerchief stuffed in a back pocket, straw hat with front and back brims folded down, a plug of chewing tobacco in a cheek, he seemed a prototypical "redneck." Still, I was surprised to hear him ask, “Where did you get that fancy New York car, boy?”

Surprised but trying to remain cool, I turned toward the voice—only to realize he was talking to Steve. In true New York fashion, Steve replied that he had bought it, probably for more than the gas station was worth. Thinking about the incident later, I recalled something I'd learned back in fifth grade: Skin color doesn't determine who's smart or dumb, nice or nasty, or athletic or clumsy—but it still meant something to small-minded or ignorant people. And color wasn't the only basis for prejudice.

Fortunately, the remainder of the trip to Kentucky passed without incident.

The Armor Officers Basic Course provided the technical and tactical foundation my classmates and I needed as new armor and cavalry lieutenants. I began to think of all the officers in the course—whether commissioned from West Point or from one of the fine senior ROTC programs across the country—as my peers. While being an Academy graduate carried some distinction, I had no interest in appearing elitist. I quickly realized that it wasn't the commissioning source that determined whether someone was mediocre or a superstar, instead it was the person.

What we all looked forward to most was leading a platoon of soldiers. The coursework was interesting, and the hands-on time with equipment was even better. The three-month course passed quickly, with full days—and occasional nights—of training. When it concluded, I packed up my Firebird with most of my worldly possessions and headed for Fort Hood, Texas, located roughly midway between Dallas and San Antonio. Texas is large, and Fort Hood was still rather remote. The drive was long and at times boring, though thankfully uneventful—no repeat of the incident in Alabama. I managed to stay near the seventy mile per hour limit, though the big V8 under the Firebird's hood begged to run free, but I resisted, wary of Texas's notorious state troopers. A couple of times, I chuckled at the thought of writing Steve to say that traveling without my Jewish buddy had helped me avoid any unpleasantness with locals. That kind of dark humor was common at the Academy and throughout Army life.

Fort Hood was huge—both in size and in its number of units. It housed III Corps, which had two armored divisions and several supporting units. The post spanned over 330 square miles—more than 200,000 acres—and had close to 30,000 soldiers assigned, along with 15,000–20,000 family members living on or near the base. Fort Hood was undoubtedly the financial powerhouse of that part of Central Texas, which caused the locals some consternation. Many viewed soldiers and their families as outsiders but acknowledged their economic importance. The adjacent town of Killeen was inexplicably a dry town where alcohol wasn't served or sold. Thirsty soldiers had either to drink on post or travel elsewhere—much to the benefit of local enforcement, which maintained lucrative speed traps east and west of the fort.

I was assigned to the 2nd Armored Division, nicknamed "Hell on Wheels." A full armored division typically comprised up to nine tank or mechanized infantry battalions, four artillery battalions, and about half a dozen various support battalions. Across the post from us was the 1st Armored Division, "Old Ironsides," identical in size and structure. At the time, I had no idea I would later serve with both divisions.

The most prevalent type of battalion in Hell on Wheels was its tank battalions, each with fifty-four tanks. These were divided into three companies of seventeen tanks each, and further divided into platoons of five tanks each. The remaining tanks were part of the battalion and company command groups. I received the assignment every armor lieutenant wanted: command of a platoon with five M60 Main Battle Tanks and a crew of twenty soldiers, including myself. To me, this small force represented the spear's tip—a formidable, mobile unit ready to defend the nation. My platoon became the pride and focus of my life. I made it my mission to learn everything about my soldiers and the equipment. Back then, the Army worked full days Monday through Friday and a half day on Saturday. Payday—always the last workday of the month—typically came with a half day off for "payday activities." This routine was sometimes disrupted by multi-day field exercises, where we camped out in the field day and night, for a couple of weeks at a time. Maneuvering, practicing attacks and defenses, building and breaching obstacles, crossing rivers and dangerous terrains, and firing the tanks' large cannons and machine guns were common activities. My job was to lead, train, and command my platoon—ensuring it maneuvered and fired effectively in tactical operations. As a twenty-two-year-old lieutenant, I found it exciting and deeply rewarding. After all, I was "playing soldier" with real tanks and weapons—and getting paid for it.

There's an old Army saying: "You'll never get rich." That was certainly true. As a new lieutenant, I received approximately $300 a month, or roughly $280 after taxes. Actually, that wasn't bad in 1968 for a single guy, but my ninety dollars per month car payment for the Firebird felt hefty. After a couple weeks living in an old, wooden World War II-era Bachelor Officer Quarters (BOQ) on post, I moved off post into a rented house with another lieutenant. It was a small, 600-square-foot home with two bedrooms in the Marlboro Heights section of Killeen, which had very modest working-class homes. This far from luxurious bungalow was better than the 10-foot-by-10-foot BOQ room, where communal latrines were at the end of the hall. Plus, having a house meant we could now socialize with friends in the evenings and on weekends after Saturday's half-day of work.

My share of the rent and utilities was another ninety dollars a month, but since I'd voluntarily moved out of on-post housing, I wasn't authorized any additional housing allowance, since space was available on Fort Hood. Routine expenses—weekly haircuts, laundry service for my freshly-laundered and starched fatigues, and gas for the car—usually totaled another forty dollars. That left me about $110 for food, personal purchases, or social activities. Like many lieutenants, I learned quickly that we often had "more month than money." The solution was to reduce expenditures for food, beverages, and social outings. The last week or so of the month usually meant lunch at the unit mess hall—about a dollar—and maybe a fast-food burger for dinner if I had any change left in my pockets.

Financial challenges notwithstanding, life was pleasant enough. Several West Point classmates were stationed at Fort Hood, and I made new friends, including my housemate, Will Lowery. He was about three years older than most of us and had been a teaching assistant before entering active duty. While I didn't realize it at the time, the three months I shared a house with Will would end up changing my life forever. More about that later.

Like most stateside posts during that era, Fort Hood was a constant churn of troops cycling in and out for deployments to Vietnam. One ubiquitous question was, "When are you shipping out?" I knew my turn would come sometime after May 1969. For my assignment, picked the same way as my branch, I selected the 2nd Battalion, 66th Armor at Fort Hood, the same unit where I'd spent a month during the summer of my junior, or Cow, year. That summer training was meant to expose us to the "real Army." But in the two intervening years, the entire 500-man battalion had turned over.

My expectations to hone my leadership and tactical skills were soon fulfilled. Much of the training we'd received to date had been theoretical or classroom-based, with a modicum of hands-on experience. Now everything was real—real dirt, real equipment that could break down if not properly maintained, and real soldiers—usually draftees—who just wanted to finish their obligation to Uncle Sam. I also commanded a range of non-commissioned officers, from junior "buck" sergeants with just a couple years in the Army to the most senior, Sergeant First Class Hector Flores, my platoon sergeant and second in command. Flores was a source of great knowledge and Army wisdom. In his fifteen years of service, Flores had seen and done much, and he dealt with many soldiers and junior officers. He knew that the better the officer, the stronger the platoon—and the better his job would be. Like all good NCOs, he took seriously the role of helping to train and develop his lieutenant.

I learned a great deal from SFC Flores on how to motivate soldiers, discipline them, and care for their well-being. The most important lesson was the immense value of the NCO Corps. At every level—whether lieutenant, captain, major, lieutenant colonel, colonel, or general—officers had NCO counterparts. A good officer respects, values, consults, and listens to their wisdom learned from experience. The officer, of course, remains ultimately responsible for leading, making decisions, and ensuring his unit's performance.

▾ ▼ ▾

Like most lieutenants, I halfheartedly complained about the endless field training when casually talking with buddies. But like most, I secretly loved it. This was exactly what I had prepared for at West Point and in the various Army schools I'd attended. Naturally, that left precious little time—not to mention little money—for a social life. Still, where there's a will, there's often a way. After nearly two months of living in the bungalow, my housemate and I were so exhausted by New Year's Eve that we didn't even bother seeking entertainment for the evening. We had spent nearly every spare moment chasing fun—house parties, social events, and the perennial pursuit of "the perfect female."

Then, just two weeks later, the most important thing that would ever happen in my life occurred. One Sunday evening I returned home after an afternoon visiting friends and saw three young ladies at our front door. They were students at Central Texas College just a few miles away. One of them knew my housemate, and after church choir practice, they had stopped by to visit him—and, by chance, the guy he was rooming with. I introduced myself and invited them in. After pleasant conversation, I learned a bit about them and, most importantly, collected their names and phone numbers.

They indicated they needed to return to campus to make curfew, so with the most gallant and debonair air I could muster, I offered a ride in my new Firebird. Two of the women accepted the offer; the third had her own ride. I hoped the allure of my lieutenant's uniform, my own home, and my flashy new car would make a favorable impression. However, as I dropped the two women off, they both seemed friendly but casually aloof, with an attractive worldly confidence. Both were very good looking, "fine Sugar Babies," as we said back then. Little could I imagine that one of them was about to change my direction and become my life partner in a union that surpassed fifty years

Field training and other duties prevented me for about a week from calling either of the young women. When I finally could, I called Nannette—the one who'd most sparked my interest with her easy smile, genial warmth, and spark of spirited nature. She asked why I hadn't called sooner, and I explained as best I could. As she would do innumerable times over the years, Nannette reminded me that she was an "Army Brat" and had "been in the Army longer than you." She understood the demands of military life. Her father, First Sergeant Dallas Owens, was a career soldier in airborne field artillery. By the time I met her, she and her family had lived in Japan, Germany, Fort Hood, and Fort Campbell, Kentucky.

That phone call led to our first date, the weeks that followed felt like something out of a movie—a war-time couple drawn together quickly, discovering each other, and growing closer by the day. It was a whirlwind romance. We spent nearly all our free time together. Before long, she even offered to teach me how to shop at the commissary and buy ingredients for filling meals, rather than living off TV dinners as I had been. Best of all, on weekends she would cook those meals in my extremely small kitchen. By then, my roommate had been ordered back to the BOQ by his commander, leaving me alone in the house and responsible for the full rent. Nannette's tutoring on shopping and skill in the kitchen were lifesavers during that time.

By April, we both knew we had found the partner we wished to share the adventure of life with. In traditional form, I needed to meet her family and formally ask permission to marry Nannette. With growing trepidation, I drove from Fort Hood to Fort Campbell, Kentucky, where her parents were stationed. My understandable apprehension was compounded when Nannette reported her parents' reactions when she told them that a young man was coming to ask for her hand. Her mother, Lillie, was reserved, wanting to meet me first. Her father, First Sergeant Owens, asked two questions: "Is he a GI?" and, upon learning I was a lieutenant, followed up with, "Is he a leg?"—Army slang for a non-paratrooper. For him, being airborne epitomized a soldier with gumption and promise versus one of the underachievers. Nannette replied that I was a Ranger qualified and scheduled to attend jump school in June. His response? "He isn't airborne yet." That vague reaction did nothing to calm my nerves as I approached Fort Campbell and what was becoming a somewhat feared encounter.

But as I'd learned time and again, confidence, faith, and confronting challenges were key. When I arrived, I found her family warm and welcoming. After the initial awkwardness, they came to see me as a sincere young man who loved Nannette and intended to take good care of her. To me, they embodied everything a good Army family should be: resilient, adaptable, worldly, and considerate. They had mastered balancing the normal pressures of domestic life with the frequent moves, differing locales, extreme working hours, and other novelties of military family life.

Back at Fort Hood, exactly 119 days after we first met, Nannette became Mrs. Larry Reginald Jordan. We had a traditional military wedding at the Division Memorial Chapel, complete with a saber arch, formed by fellow officers, for the newly wedded couple to walk under. I was fortunate to meet more of her family for the first time. My best man was Lieutenant Lew Jones, a friend I'd met at Fort Knox who was now in my battalion. Lew, a Morgan State University graduate, and cool guy, would turn out to be the first of many Morgan State alums who became very close friends and de facto extended family.

Within weeks, Second Lieutenant Jordan and Mrs. Jordan departed Fort Hood to head to Fort Benning, Georgia, where I was slated to attend Airborne School. After that, I would deploy to Vietnam. Completing Airborne School would please First Sergeant Owens and earn me another badge of proficiency and success: silver jump wings pinned on my uniform. Soldiers often checked each other's uniforms, scanning for signs of experience and skill: Ranger tabs, airborne wings, marksmanship badges, and the coveted Combat Infantryman's Badge (CIB), which denoted combat service in an infantry role. As an armor and cavalry officer, I wouldn't be authorized a CIB, unless I was assigned to an infantry position in Vietnam (something I couldn't have predicted, but did happen).

Our trip from Fort Hood to Fort Benning marked the first of what would eventually be twenty-three moves for the Jordan family—something neither of us could have envisioned. We hitched a U-Haul trailer to the Firebird, packed up our worldly goods (which didn't even fill the trailer) and set off. Since the Firebird lacked air conditioning, we did most of the driving by night to stay cool and avoid traffic. When I'd bought the car, I reasoned that air-conditioning would only sap horsepower in my small sports coupe. The most memorable moment of the trip came on July 20,

1969: the *Apollo 11* moon landing. We watched the event unfold on a small motel television, and I was entranced. I had long been a space fanatic. But Nannette, practical as ever, insisted we hit the road to make the most of the cool evening travel time. I reluctantly agreed, comforted by the fact that we could listen to the broadcasting on the car radio. Somewhere outside of Memphis, we heard Neil Armstrong's famous words, "That's one small step for man, one giant leap for mankind." In many respects, Nannette and I were taking our own leap into the unknown—though we had no idea where it would lead.

When we arrived at Fort Benning, being a married officer attending a short course didn't qualify me for on-post housing. Like many other married lieutenants, I rented a room at the Camelia Motel, an old, sprawling complex immediately outside the main gate. It was close, available, and—most importantly—inexpensive. That mattered, since even with my promotion to first lieutenant two days before the course started (and the accompanying pay bump), paying to support two people was a new challenge.

Jump School lasted three weeks. While several West Point classmates were in the class, most students were junior enlisted soldiers preparing for assignment to airborne units like the 82nd Airborne at Fort Bragg and the 101st in Vietnam. There were also other lieutenants and even a captain, many from non-Academy backgrounds—a reminder that, going forward, West Point graduates would make up less than 20 percent of the officer corps.

In those days, the unit patches and other insignia sewn onto fatigue uniforms were full color—unlike the subdued black-and-green versions adopted in later years. Having just left the 2nd Armored Division and not yet reported to my new unit in Vietnam, I wore the armored division patch: a colorful triangular patch of yellow, red, and blue, featuring a tank track, cannon, lightning bolt, and read HELL ON WHEELS embroidered beneath. While eye-catching, it also made me a prime target at Airborne

School not because of my race but because of the patch's placement. It was certain to draw attention and scorn because the 2nd Armored Division uniquely wore its patch on the left chest rather than the left sleeve. The result was good-natured hazing, in the form of extra pushups and other exercises administered by the NCO trainers, or "Black Hats," so called because of the black ball caps they wore. In a class mostly made up of second lieutenants and privates, any distinguishing feature stood out. My West Point classmates and I were also newly minted first lieutenants, which meant replacing our gold bars on our collars with silver ones.

Being a first lieutenant brought a small bump in pay but, more importantly, a distinct bump in respect. Second lieutenants were expected to know little and perform accordingly, whereas first lieutenants were expected to have more experience and maturity. At Jump School, this status caught the attention of the Black Hats. I learned to dread the call, "Where is that hardcore tanker lieutenant?"—usually followed by, "Get that hardcore lieutenant to demonstrate how to properly do the pushup!" Though it felt like harassment or insubordinate punishment at the time, it was a sign of respect. The Black Hats appreciated officers who volunteered for Airborne School despite not being assigned to airborne units. Putting officers front and center also modeled to enlisted soldiers the deeply ingrained Army tradition that their leaders were sharing the same hardships as their soldiers. The officers follow such practices as not eating until all soldiers are fed in the field, not leaving for home until all soldiers ended their workday, and doing all the training and being proficient in the same things the soldiers are required to do.

▼ ▼ ▼

After two weeks of learning to handle the pieces of parachute equipment, practicing proper technique of jumping from a plane's cabin, and mastering landing skills to land without serious injury, the culmination was five real-world parachute jumps. At that time, the Jump School was supported by the Air National Guard using C119 cargo planes, nicknamed "Flying Boxcars." Though reliable, these ancient birds—from a World War II design and used in the Korean War—were somewhat smelly, rough riding, and unsettling for first-time jumpers. When the side door slid open at 1,150 feet, objects on the ground appeared tiny, but low enough to only take a few seconds to reach the ground at tremendous speed if the parachute didn't deploy.

On my fifth and final qualifying jump, I decided to highlight and honor my armor and cavalry lineage even more than the bright patch on my chest did. Paratroopers traditionally wore highly polished, lace-up boots, preferably Corcoran brand, that were cut high and supposedly helped stabilize and prevent ankle injuries during parachute landing (and, perhaps more importantly, looked sharp). Anything else was taboo. But tankers and cavalrymen—especially NCOs and cocky young officers—often wore distinctive "tanker boots," which had no laces and instead fastened with leather straps with buckles that wrapped dashingly around the boots to hold them on snugly. These were highly polished and had the perfect jaunty look to accompany a swaggering tanker or cavalry trooper. For my final jump, I donned my best pair of tanker boots. The first Black Hat to see them was incredulous. Finally regaining the ability to speak, he pointed them out to the other NCOs, shouting "What the hell are you doing in those orthopedic boots, lieutenant?" I told him that I wanted to show how Hell on Wheels troopers jumped. After some consternation, they let me, this crazy lieutenant, proceed, and break his ankles if he wanted. I promised that I'd land safely and "motor off the drop zone under my own power." I did—and afterwards earned respect from my classmates. It was another lesson: Within reason, it's okay, sometimes even advantageous, to display a little dash and flair.

We had about a week before I was scheduled to deploy to Vietnam. Nannette and I tried to put that thought aside. She planned to stay with her mother at Fort Campbell while I was gone. Her dad was slated for deployment for Korea later that year, and most families weren't permitted to accompany soldiers there due to a lack of facilities. Mother and daughter would be waiting wives together. Fortunately, with many soldiers being assigned to Vietnam, the Army allowed waiting wives, like Nannette and her mother, to remain in base housing while their husband was deployed—something that turned out to be a blessing when Nannette discovered she was pregnant with our first child. It was also helpful that she had access to the Army hospital and medical care. By having her mother and free medical care close by, it gave us peace of mind.

As we prepared for my year-long deployment, we both realized events since January were a whirlwind—like a dream running at triple speed. We feared the next year would be the opposite, dragging by painfully slowly, separated by thousands of miles and, in my case, serving in a war zone. Early on, I had once suggested we wait until I returned from Vietnam to get married, thinking it unfair to leave a new bride behind while going to war and perhaps never returning. But Nannette would not have it, insisting that I would come back safely if it was God's will. Now, those thoughts crept back as we faced an unexpected pregnancy and the realities of military life. Still, her strength and resiliency amazed me. Watching her handle it all taught me something no classroom or Army school ever could: how strong and tough military wives—like Nannette—had to be. She'd grown up in the Army and knew exactly what she was signing up for. She knew that mothers and other wives had to manage. Many times, over the five decades of marriage to follow, I would tell her that journey would not have been worth it without her at my side.

On a bright late June day, Nannette and her mother drove me to the airport. Wearing my khaki uniform, I boarded a plane bound for Oakland, California. Twenty-four hours later—after stops in Hawaii and

Japan—I landed at Tan Son Nhut, South Vietnam. Just like my arrival in New York City years earlier, I felt like I was stepping into a completely different world. But unlike New York, this world came with real risks of lethal danger that could easily occur.

CHAPTER 5

Vietnam

In some ways, the unit in Vietnam was like Fort Hood, with people constantly arriving and departing as their one-year tours began or ended. I learned at the in-processing center that no armor lieutenant slots were open, but lots of Infantry replacements were in high demand. I was to join the 1st Infantry Division, nicknamed the "Big Red One" for its status as the Army's oldest division and the red numeral "1" on its otherwise plain olive drab patch. Not only was it the oldest, but also one of the most highly decorated units from World War I and World War II. It was now fighting in Vietnam. It also had a cavalry squadron as part of it, so my skills could be applied. However, that wasn't to be. Upon arrival, I was informed the squadron had a full complement of officers, and I would be assigned to an Infantry battalion—always in need of lieutenants. Some of my Infantry Classmates from the Academy might have been a bit envious, having been assigned to less famed divisions. Here I was, a tanker, receiving a prestigious Infantry posting. I even joked that it was logical because the Army wanted the best lieutenants in place to win this war.

After a couple days of in-country indoctrination at the in-processing center, I was sent to division headquarters at Di An. From there, I flew by helicopter to the 1st Brigade base camp near Dau Tieng, a few kilometers southwest of the Michelin Rubber Plantation—the largest in the country, and some say the largest in the world. When I arrived at the rear headquarters of my assigned unit, 2nd Battalion, 2nd Infantry (Mechanized), I had to wait several hours for transport to my company in the field. During that time, I remained at the rear operations center, manned by a captain and several privates and junior sergeants monitoring radios. The radios blared with conversations from units engaged in a sharp firefight. Listening to the urgent but controlled voices, using phrases and code words I had learned at Fort Knox and elsewhere, I realized this was not training—this was real. One of the radio operators, a twenty-something-year-old sergeant, winked and said, "You arrived just in time, lieutenant. They're going to need replacement platoon leaders." Dark humor like this was common as far back as West Point, but this time it carried a different weight.

Finally, I was transported to Charlie Company (C Company), 2nd Battalion, 2nd Infantry. A real surprise awaited me: The company commander was a brand-new captain. He was a 1967 graduate who had been in the cadet company next to mine at West Point. Even more surprising—and infinitely less probable—the other two platoon leaders were also my classmates, having arrived about three weeks earlier. The statistical probability of that was so small as to be unreal. We later discovered the reason: the battalion commander, a lieutenant colonel, had been an instructor at West Point and had taught the Class of 1968 during our Plebe year. Knowing that three graduates from that class were headed to the division, he arranged for all of us to join his battalion. He then assigned us to the same company, under a commander we might know. The experiment worked superbly—there was instant trust and bonding, and the company became a smooth, effective fighting machine.

Although I was serving in an Infantry battalion, I felt at home tactically as prepared as my two Infantry classmates. Ours was one of only two mechanized infantry battalions in the division. The vehicles were M113 armored personnel carriers. Though they ran on tracks, they were not tanks. They offered modest protection with their aluminum armor and featured a .50-caliber machine gun, as well as lighter M60s. The carriers also hauled large amounts of ammunition, water, rations, and supplies—freeing soldiers from having to carry everything on their backs. Most operations were still conducted on foot, moving through jungle, bamboo, or chest-high tall grass, while the vehicles followed at a distance to provide support. Approaching potential enemy positions on foot was safer and allowed better visibility and the ability to approach undetected.

In that area of Vietnam, combat typically consisted of small-unit actions, with occasional large units slugging it out. The enemy rarely engaged in large battles unless the terrain, weather, or circumstances offered them some advantage to counter the U.S. firepower, aviation, and artillery. Still, enemy battalions and regiments were present and would mass for attacks if a lucrative opportunity arose. These smaller engagements gave platoon leaders the opportunity to demonstrate how they responded to enemy and employed their assets. The next four months saw several sharp engagements, many around the edges of the Michelin Rubber Plantation. The plantation wasn't operational due to the war, though a representative of the Michelin family and company periodically visited to inspect the remaining facilities.

Prior to the war, the plantation had been organized like a small village. Shaped roughly like a diamond standing on point, it was divided into four quarters by intersecting east-west and north-south roads. Each area included processing buildings, administrative offices, and other structures. The plantation also had a few sleeping quarters for workers, communal kitchens, and similar facilities. All of these were abandoned and in severe disrepair—many reduced to just walls or foundations. Part of the battalion's mission was to keep the Michelin clear of Viet Cong (VC) and North Vietnamese Army (NVA) elements.

One of the unit's primary missions at the time was to provide security for what were known as Rome plows. These were large specialty blades mounted on the front of huge Caterpillar bulldozers operated by the Army engineer troops. The blades were manufactured by the Rome Company, originally based in Rome, Georgia. The entire machine—bulldozer with an armored cage over the driver and blade—became known as a Rome plow, and it was used to cut large swaths through the jungle and vegetation. The resulting clearings were often several football fields wide and could stretch for miles. The cleared areas served two purposes: For one, they denied the enemy cover and concealment. Secondly, they enabled ground radar to detect and target enemy movement at night or whenever U.S. forces lacked direct visual observation. For the mechanized unit providing security, this also offered opportunities to locate and engage enemy soldiers hiding in camps or bunkers within the cleared zones.

The enemy hated the plows not only because of their clearing power but also because they believed the heavy machines could collapse the many shallow tunnels they often constructed and used. As a result, the Rome plows and the engineer soldiers operating them were prime targets—typically attacked with rocket-propelled grenades and rifle fire. Occasionally, the enemy tried to ambush the plows by planting mines and boobytraps along their path, and positioning soldiers with machine guns and rifles along the flanks. Charlie Company's mission was to escort and protect about a dozen plows as they worked. Encounters with Viet Cong or North Vietnamese forces would often end quickly, given the American superiority in firepower and protection. Ambushes, if not detected before the plows detonated a mine or boobytrap was triggered, could take considerable fighting to eliminate the threat. Still, it was usually a one-sided affair, with U.S. direct firepower, artillery, and air support eventually overwhelming the enemy.

At night, our plows would return to a somewhat spartan, circular field base 200 yards in diameter. It was behind a 6-foot-high earthen berm, pushed up by bulldozers. Outside of that lay strands of barbed wire and razor wire to discourage intruders. Positioned among the wire were mines and claymore munitions—lethal directional explosives that could blast hundreds of ball bearings toward the enemy on command.

Inside the berm, our M113 armored personnel carriers formed a defensive ring, each with their array of machine guns and armed soldiers. In the middle of the base camp usually stood two 105mm howitzers and four heavy mortars, capable of firing defensive fires. Those could be high explosive rounds, or if necessary, illumination rounds, which when fired, lofted a bright, burning flare that floated down slowly on a small parachute, illuminating the battlefield, and denying the enemy the cover of darkness. Each night, once the camp entrance was sealed with a portable metal barricade covered in barbed wire, and protected by mines, there was no easy access. Still, the enemy was so determined to disrupt or destroy the plows, they attempted to overrun the base camp several times in the darkness. Some fights lasted hours, with VC and NVA forces sustaining heavy casualties. Due to the armored vehicle protection and available firepower, Charlie Company suffered few wounds.

My closest call came during one nighttime attack. I was standing at the open door of the rear ramp of my vehicle when the first mortar rounds hit. One landed close enough to blow me through the door onto the vehicle's floor. I didn't receive a scratch, which astonished both me and my crew. Later, in a bit of false bravado, I told them, "It wasn't my time to go, and the VC who can take me out hasn't been born." Inwardly I knew that wasn't the case.

Four months after arriving, I was reassigned to the battalion's Reconnaissance Platoon—a plum assignment reserved for experienced lieutenants who'd demonstrated the ability to operate without close supervision. The platoon operated much like a separate company. I exchanged the four vehicles for six M113s. These were seriously different. Instead of the single .50-caliber heavy machine gun, five of the vehicles had three guns. One gun was operated by the vehicle commander and the other two were mounted at the rear corners either side. These were called

ACAVs, or armored cavalry vehicles. That gun arrangement allowed at least two guns to fire in any direction where a threat appeared. All three could fire at targets to the front.

The sixth vehicle had a single heavy machine gun but also carried a 106mm recoilless rifle mounted on a swivel on the vehicle's right side—a weapon equivalent to a tank cannon but without the sophisticated sighting and aiming systems. It used simple telescopic sights, which were effective at the anticipated ranges. Now, I felt that I had some real firepower. And there was more.

I also had a three-vehicle section of tracked throwers—modified M113s nicknamed "Zippos." Inside were two round 300-gallon pressurized tanks—one with compressed air, the other jellied gasoline. In other words, it was ground-launched napalm, able to shoot a stream of the burning mixture up to one hundred yards. The soldiers riding these Zippos were considered a bit crazy by others. Who else would wish to ride around atop that load? Yet those soldiers were as careful and courageous as the rest and were greatly appreciated when more firepower was needed.

The jellied gasoline was also used to fill 55-gallon metal barrels emplaced around the outer protective berms and within the razor wire circling base camps. Known as "foo gas," these barrels could be detonated in the event of an attack, with devastating results. Since the napalm tended to degrade within a couple weeks, a few barrels were detonated every few days and replaced. These detonations were timed as nightfall demonstrations—just when enemy soldiers might begin to move, observe, and probe friendly positions.

Captured enemy soldiers later confirmed that the specular flaming eruptions, along with Claymore blasts and the coordinated practice weapons fire, known as "Mad Minutes," dampened the desire to attack the camps.

Two months after joining the Recon Platoon, I became involved in the largest battle I experienced during the entire tour. My mission was to interdict enemy movement through the Michelin Plantation. The rest of our battalion was operating farther south, but with the addition of three M48 main battle tanks and three heavy mortars, it was believed that our platoon would have the ability to secure the area without assistance. Two ACAVs were away being retrofitted with belly armor, but both the battalion commander and I felt the available force would suffice.

On November 22, 1969, the members of this small but potent force heard and felt tremendous explosions several miles to the southwest. Huge clouds of smoke and dust rose from that direction. It quickly became evident that a B52 "Arc Light" bombing strike was in progress. Platoon members joked that someone several miles away was having a bad day—and we were happy it wasn't us. Twenty-four hours later, a radio call was received directing us to move toward the site of the air strike to assist a battalion in trouble. I didn't have a clear picture of what to expect.

With the four vehicles available, I elected to take the three tanks and leave the Zippos behind to protect the mortars and base. The flame tracks wouldn't have been able to move fast enough anyway. While headed overland to the objective, roughly 14 miles away, one tank broke down, and another threw a track. Rather than wait the hour or more needed to replaced it, I left the third fully operational tank to guard the disabled pair while repairs were made.

The cryptic radio call from the battalion command center hadn't described the situation in any detail. In those the days, most radio communications weren't encrypted, and the enemy was known sometimes to intercept radio messages. The people at the command post didn't want to reveal how desperate a situation that U.S. battalion was in. Still, the implications were clear. If an entire U.S. battalion of close to 600 men couldn't handle the fight, I wondered what my small but potent force would be asked to do.

When we arrived, the perilous situation was obvious. The B52 "Arc Light" bombing strike had obliterated one-third of a large bunker complex shaped like a three-leaf clover. One "leaf" was completely destroyed. After the B52s dropped dozens of 1,000-pound bombs, the ground had been pulverized, turning the soil into fine light brown power. Huge craters, about 6 feet deep, covered the landscape, each had quickly filled with a foot of crystal blue water from the high water table.

Beyond the craters and ruins of the destroyed bunkers lay the two other leaf-shaped bunkers clusters. Though untouched, they were filled with shaken and extremely perturbed NVA soldiers. They amounted to about two-thirds of a battalion, with one-third lost to the air strike.

The U.S. infantry battalion had deployed a two-company force from the south to clear the enemy from the bunkers. Inexplicably, they advanced through the bomb craters and were pinned down by heavy fire from the untouched bunker positions. Unable to advance or retreat, the pinned force requested assistance. A third company from the battalion was inserted by helicopter to the north of the complex to attack from that direction. But they too ran into the untouched bunkers and were halted by enemy fire. The situation was getting out of hand as the entrenched enemy was slowly whittling away the attacking Americans. The appearance of the four Recon Platoon tracks both boosted the morale of the beleaguered infantrymen and turned the tide.

Driving directly at the bunkers, the platoon silenced them one by one with heavy machine gun fire and hand grenades. The process was slow. When the remaining bunkers opened fire with rocket propelled grenades, I began destroying them from a distance using the 106mm recoilless rifle. After several bunkers were destroyed that way, the remaining NVA, realizing their fate if staying in place, evacuated the bunkers and fled.

Exposed in the open, many were dispatched by heavy machine gun fire. More than one hundred enemy fighters were killed in that fight. Although his troops had been pinned down until the vehicles arrived, the infantry battalion commander claimed the majority of these casualties as his battalion's work. I did not complain. The assistant division commander, a brigadier general, had observed everything from his helicopter overhead. He knew the truth.

That day in November 1969, I only did what I thought was my duty—to fight, destroy the enemy, and bring Americans back home alive and well. I didn't feel that I had done anything extraordinary. But I made certain all of my soldiers were decorated for their bravery and performance in battle.

Later, I was awarded the Silver Star Medal for gallantry in action, leading a platoon to assist a pinned down battalion. The citation reads, in part:

> For gallantry in action while engaged in military operations involving armed conflict with an armed hostile force in the Republic of Vietnam: On this date, Lieutenant Jordan was serving with his mechanized unit on a reinforcement mission in the Michelin Rubber Plantation. Another friendly unit had attacked an enemy base camp and had been pinned down, Lieutenant Jordan's unit maneuvered with their tracks to the point of contact. When he reached his beleaguered comrades, Lieutenant Jordan brought the .50 caliber machine guns into operation and instantly placed devastating accurate fire upon the insurgent positions. With tactical deliberation, he then led his unit on . . . an assault of the positions. Lieutenant Jordan risked his life as he personally placed suppressive machinegun fire from his track and . . . directed the destruction of enemy bunkers with hand grenades. Continuing to disregard the hostile fire raking the area, he coordinated the assault of the enemy positions . . . until many of the insurgent strongholds were silenced. When the last remaining insurgent position began placing heavy rocket-propelled grenade fire on the friendly unit, Lieutenant Jordan . . . directed its destruction with recoilless rifle fire.

That day reinforced another lesson. Don't worry about who gets the credit. Just do the job. The facts and truth will prevail—and knowing you did a good job is what's most important. I didn't think of my performance that day as particularly heroic. It was just what it was.

▼ ▼ ▼

After that tough battle, I looked forward to the holidays—my first as a married man. I remained in the Michelin Plantation, continuing the independent operations while the rest of the battalion worked the area to the south. The holidays traditionally witnessed an informal truce, with both sides suspending offensive operations from a few days before Christmas through New Year's. U.S. forces conducted defensive ambush operations to prevent attacks from any enemy forces that chose to disregard the truce.

On Christmas Eve 1969, I found myself quietly reflecting on what it meant to be in combat during the holy season. I had deployed a squad-sized defensive ambush party as usual. But as I looked up at the clear, star-studded sky, my thoughts drifted 1,000 miles away. Suddenly, one of the booby traps employed around the ambush site detonated. The squad on the ground didn't see or detect any enemy and there was no return fire—the standard practice when either side was caught in an ambush. To prevent disclosing their presence or location, the American squad on the ground held its fire. The enemy would think it had triggered an unmanned booby trap or stumbled into a mine field. I called in artillery fire around the ambush site to destroy any survivors. No further movement was observed or heard that night.

That evening was introspective. The idea of being involved in combat action on such a sacred evening weighed on me. A few hours later, following a restless night, I awoke to urgent calls from the battalion headquarters asking for a report on the ambush action. I radioed the squad leader seeking information on possible enemy casualties, though I felt saddened by the prospect of having taken a life on Christ's birthday.

The ambush leader responded with a chuckle: "Sir, we got a whole family of them." Puzzled, I wondered whether civilians had somehow wandered into the ambush that night and asked if any had been carrying weapons. "Most had some wicked-looking tusks," he said.

It turned out to be a family of large wild boars. In truth, most in the platoon and battalion headquarters probably felt a sense of relief. While everyone took the mission of engaging and destroying the enemy seriously and carried it out aggressively, they were grateful not to have done so on Christmas. The boar meat was donated to an appreciative local Vietnamese village, which felt fitting for spreading the holiday spirit.

A couple of weeks later, something even more personally significant took place. The long-awaited moment finally came on January 16, 1970, while the platoon was out on another mission north of the Michelin Plantation. We were operating near a small, abandoned rubber plantation and a nearby Special Forces camp when, the next day, word reached me through a Red Cross message: Larry Reginald Jordan Jr. had been born.

Surprised, I took special note of the name. Before departing for Vietnam—and in letters exchanged with Nannette since—we had vigorously discussed possible baby names at length. This was before the days when ultrasounds could definitively determine gender, so we had made lists of possibilities for both a boy and a girl. I had thought that Phillip would be a good name for a boy, even though no one in either family had ever been named Phillip. I had simply assumed—without any confirmation from Nannette—that if it were a boy, he'd be named Phillip. Obviously, my independent and strong-willed wife had other ideas.

I was proud of her for standing her ground and inwardly surprised at how honored I felt to have a son named after me.

I also recalled something I hadn't thought about in years—my own father's name was Napoleon Bonaparte Jordan. That unusual name had led to a humorous development back when my parents were dating. Dorothy hadn't found the name "Napoleon" particularly attractive and was concerned that her friends might tease her new boyfriend about it. She decided "Larry" was more suitable—and more likable. She introduced him that way, claiming it was his nickname. For years afterward, her old friends continued to call him Larry.

The real Larry Jordan, me, couldn't help but wonder whether there had been any debate over the naming of Napoleon and Dorothy's firstborn. I was fairly sure I wouldn't have liked being named Napoleon, imagining the endless teasing I'd have endured in school.

I also wondered why my father's parents had chosen that name. Perhaps they thought their son might grow up to be a military man or a leader. But as poor farmers from Arkansas and Oklahoma, there really wasn't much chance of that. But then again, when I was born, did anyone foresee my one day graduating from West Point and having a career in the Army?

That evening I glanced over a small Bible with a black leather cover embossed with my name. Each Cadet had received one from a national religious organization. I had brought it to Vietnam, trusting that God would bring both it and me safely to Nannette. At the time, I hadn't even considered the possibility that taking this particular Bible to combat would become a family tradition. In the coming years, I would take it with me when serving along the Iron Curtain, in Iraq during Operation Desert Storm, and later to Bosnia and Kosovo. My newborn son would one day take it with him to Iraq and Afghanistan—and Larry Jr.'s son, my future grandson, would take it into Afghanistan as well.

A couple more months of operations passed until the Army decided to redeploy the entire division to its home base at Fort Riley, Kansas. Soldiers who had completed ten months of their tour by the deployment date were allowed to end their service early and return home with the Big Red One. I missed the cutoff by just two weeks and would remain in Vietnam until late June, instead of heading home in April with the lucky ones.

Then, an unusual and unexpected opportunity arose. The Commander of Military Assistance Command Vietnam (MACV)—second-ranking officer in all of South Vietnam—was seeking a new aide-de-camp. That position, typically given to a junior officer, involved arranging schedules, planning trips, and managing the general's daily affairs. It was an opportunity to observe and learn from a senior leader.

I was nominated by the division. It was the kind of assignment that could truly advance a career—one of those top shelf opportunities. I was plucked from the field by helicopter and flown to headquarters near Saigon. It felt like landing on a different planet. The headquarters was a large, air-conditioned complex of aluminum and glass. It was quickly decided that my worn field uniform wasn't appropriate for the interview, so I was issued a new set of jungle fatigues—and a kind Red Cross volunteer sewed on the patches.

The interview went well. The general was particularly interested in my field experience and how I had earned the Silver Star award. After describing the job duties, he asked how soon I could begin—only then remembering to ask when my tour ended. That was important because he needed someone for at least six to nine months.

I reluctantly informed him that I was scheduled to leave before the assignment would begin. When asked if I'd consider extending my tour by several months, I declined—explaining that a wife and newborn child were waiting for me at home. Faith, Family, Friends, and Flag may all be guiding principles, but I couldn't place Flag before Family for the sake of career. The general said he understood—and agreed. I left believing that I had done the right thing. I didn't realize then that I would need to make the same decision again years in the future.

It had been nearly a year apart from Nannette. In those days—before email, texting, cell phones, or video calls—there were only two ways to stay in touch with families and loved ones: writing letters and hoping for a rare phone call. I wrote often to my wife and eagerly awaited her letters. Like everyone else, I anxiously awaited mail call, which was when letters were distributed by the First Sergeant or unit clerk. For a few brief moments, a letter could transport you back to the people you loved and missed—but only temporarily. Once the reading was done, the loneliness and missing returned.

If a soldier was fortunate enough to pass through one of the larger permanent bases with access to what passed for military phone service, they might be able to schedule a MARS call. Civilian phone service wasn't available. MARS—short for Military Auxiliary Radio System—linked the Army's phone system in Vietnam to amateur radio operators in the U.S., who then connected the call to the stateside civilian phone system.

If lucky, a soldier put their name on a long waitlist for a ten-minute call. Demand was high, so it was best to reserve a slot as far in advance as possible and inform loved ones at home by mail of the anticipated time—helping to avoid missing connections if the person wasn't home. Phone operators back in the States often called hours ahead to let someone know a MARS call was coming. This process was further complicated by the seven- to eleven-hour time difference between Vietnam and the U.S., depending on the destination, as well as the crossing of the International Date Line. MARS calls often had radio interference or issues with the radio-to-telephone interface. The amateur radio operator facilitating the call would often have to relay messages between the two parties. It wasn't uncommon for them to say something like, "She says that she loves you." Those calls were not entirely romantic but still received with heartfelt appreciation.

The absolute highlight of that year was the wonderful week spent together in Hawaii on R&R—Rest and Relaxation. This was the one week of leave each soldier received during the combat tour. Married soldiers often met their spouses—if they could afford to fly them to Hawaii. Single soldiers, or those not meeting a spouse could travel to destinations like Bangkok, Thailand; Taipei, Taiwan; Sidney, Australia; or a few other locations, depending on available travel arrangements.

In March 1970, I reunited with Nannette in Hawaii—and met Larry Jr. for the very first time.

Once the Big Red One departed Vietnam in April 1970, I was reassigned to a separate cavalry squadron operating in the north—between Pleiku and the South China Sea. Enemy activity was much lower than what I'd seen previously, but the threat remained.

Two days after arriving, I mentioned to the squadron commander that the bunkers seemed inadequate for protection against mortar or rocket attacks. He replied that they hadn't heard a shot fired in anger for some time and considered the bunkers sufficient. The very next night, the camp came under heavy shelling, and enemy rifle fire was heard as the VC probed the perimeter. Soldiers of the new unit crowded into the small bunkers.

I joined them—and asked how many had brought their weapons. Shockingly, half had left their weapons. I ordered them to retrieve the weapons they had left behind and return. I concluded that this wasn't a well-trained or disciplined unit. They had operated independently for a considerable time and hadn't sustained enemy contact. I was thankful I'd only be there a few months.

▼▼▼

Two memorable things occurred during those months. First, I was promoted from Lieutenant Jordan to Captain Jordan. It was on the exact second anniversary of my commissioning. Every member of the West Point Class of 1968 was promoted that day. The exigencies of the war had shortened the promotion timeline from second lieutenant to first lieutenant to just one year and from first lieutenant to captain also one year—a welcome change in our eyes.

What we didn't realize then was that once the war slowed, so would promotions. My year group would go on to spend an average of ten years as captains, waiting for promotions to return to normal after the drawdown that followed peace. At the time, though, none of that was known—or mattered—to me. I was now eligible to command a company when I returned to the States in just a few weeks.

Second, a new Signal Corps officer arrived to serve as the squadron's communications officer. It was Captain Dan Robinson—my former roommate at West Point. Once again, the laws of probability were turned on their head. For my final month in Vietnam, I shared sleeping quarters with the same friend I had roomed with nearly two years at the Academy.

For the first time in a long while, my thoughts returned to the Academy. Dan and I talked late into many nights, picking up right where we'd left off. It felt as if we'd been together without interruption since our cadet days. I would come to experience that same phenomenon with good friends many times in the coming years—ten minutes together, and you were mentally back where you were last.

Occasionally, I thought about my father-in-law, First Sergeant Owens, who was then deployed in Korea, having served in Vietnam three years earlier. I thought about my recently expanded family—my wife and mother-in-law—both waiting for husbands to return from distant lands.

Since the wedding, I'd gotten to know "Top Owens" better. "Top" was the informal term soldiers used for the First Sergeant—an old Army term of respect, for the unit's senior enlisted soldier. I usually referred to Top Owens as the "Master Blaster," in deference to the Master Parachutist wings he wore instead of the standard Jump School wings. Those wings marked him as a seasoned Soldier-master, the individual in each aircraft that supervised the preparation for and conduct of the jump. I smiled, recalling some of his Vietnam stories. Now, I had my own ones to tell.

The trip home was bittersweet. While the reunion with Nannette and Larry Jr. was joyful, the nation's reception was far less so. America did not welcome home its returning veterans who had performed their military obligation or duty. Many had turned against the U.S. involvement in Vietnam as the casualties mounted. It was not like the homecoming troops would receive two decades later, returning from Operation Desert Storm.

Still, my reunion with Nannette and Larry Jr. was joyful. It felt like one part of my long road march had ended—and another was just beginning—and it had.

CHAPTER 6

Back in the World

While in Vietnam, soldiers often referred to home as "The World." Coming home was like entering a new World. The three weeks following the return from Vietnam were both blissful and hectic. Amid the joy of being with my wife and son, there was time to bond with my new extended family. Nannette was the second of five siblings, and the only girl. While growing my relationship with Nannette's mom and dad, I had my first chance to engage her four brothers. The eldest, James, was one year older than Nannette and had completed his tour in the Army. Nolan, a couple years younger than her, was headed to the Air Force. The two younger brothers, Dallas Jr. and Keith, were about twelve years and four years old at the time, while Nannette was twenty-one years old. The age difference between her and the two youngest placed her as more of a second mother than sister or playmate. Dallas was fascinated with the Firebird, which Nannette had driven the past year, and Keith was a typical, energetic boy. I tended to view both as the slightly bothersome but otherwise tolerable younger brothers I never had. Having only the one sister, six years my junior, I found the constant bustle of a large family to be a new experience.

After a week at Fort Campbell, we headed to the nation's capital to visit the Army's Personnel Center and talk with the Armor Branch manager for captains. I had hit upon the idea of requesting to go to Fort

Knox rather than Fort Riley, which sounded like it was on the edge of nowhere. It actually wasn't, but those who had been assigned at Fort Riley joked that "you could see it from there." I had never been to Fort Riley but had seen rural Kansas—it did not seem like a place I wanted to go. The rationale I planned to present was that I could get a company command at Fort Knox, then remain to attend the Captains Advance Course, the counterpart to my earlier Lieutenant Basic Course. It made great sense to me, plus I was familiar with Fort Knox, and with Louisville, a few miles to the north. After explaining my proposal to the unimpressed major who worked up assignments for all armor captains in the Army, he simply said "You want to go the Fort Riley." It wasn't a suggestion, or recommendation, but an order. Something about the way he said it caused me to think there was something more, but I dared not ask after his firm shootdown of the proposal. The lesson learned was along the lines of the old saying, "Nothing ventured, nothing gained." I didn't think I had gained anything but would find out later I was wrong. What was immediately understood was that the needs of the Army came before personal desire. This was a critical step along the path to selfless service.

Once business finished in Washington, DC, I was eager to leave the maddening traffic. The next leg of the trip was to New Haven, Connecticut, to visit my mother-in-law's only sister, Nannette's Aunt Marge. Aunt Marge, her husband, Bill, and their children, Charlotte and Carlos, immediately made me feel like part of the family. I was beginning to enjoy the flurry of news, upbeat tempo, and gregarious nature of a large family.

The following week, after another quick stop at Fort Campbell, we visited my mother and sister in Kansas City before driving to Fort Riley to report to the new unit, the 1st Battalion, 63rd Armor. Located a couple of hours west of Kansas City, Fort Riley was in rural Kansas as I had been led to believe. Kansas State University was located some miles to the east, in Manhattan, Kansas. The town of Manhattan certainly didn't compare to its New York namesake, and most soldiers and families simply shopped and visited Junction City, just outside the post gate. Much like Killeen and Fort Hood, Junction City largely owed its livelihood to Fort Riley. It was a typical Army town, with car dealerships, stores and shops, and, of course, places that catered to the entertainment of thousands of young soldiers.

The post itself had perhaps 30 percent of the population of Fort Hood and was half the size. Still, there was a large maneuver area, and several ranges that accommodated everything from rifles and pistols to tanks and artillery. Much had undoubtedly changed since 1905, when the 9th Cavalry troopers stationed there were assigned horses for transportation. Though a few old stone and brick structures from that era remained, motor pools filled with tanks and other equipment had replaced the stables for the cavalry mounts. One thing that did remain to remind all of Fort Riley's frontier heritage was a small herd of buffalo in a large pen located on the installation, preserved from the days of Camp Riley, and maintained under an agreement between the Army and the Department of the Interior. My introduction to Fort Riley was long before my interest in genealogy and attempts to research the family history. It would be many years before I learned that sixty-five years earlier my Great-Uncle Joe had received orders to the same post.

We arrived at Fort Riley on a Friday. Upon reporting to the battalion commander, Lieutenant Colonel Tom Kelly, I was informed that no family housing was available on post; I should get my family settled and return the following Monday morning. Since no rentals were available off post either, that meant taking the family back to my mother's house in Kansas

City. Kelly also added that he had seven extra captains in the battalion above his authorization—I was number eight. The steady return of officers from Vietnam, and the drawdown of forces there, resulted in extra officers being assigned to battalions. The scramble to secure a company command assignment, key to progressing further, was immense. It was hinted that I might be assigned as the battalion adjutant, a staff job involving paper pushing. That was discouraging and cast added gloom to the prospect of being separated from the family after so recently reuniting.

Lieutenant Colonel Kelly was a brisk but amiable Irishman with slightly red hair who had graduated from Notre Dame. He had a temper that flared at poor performance or those who seemed less than professional. Years later, as a three-star lieutenant general and the operations officer on the Joint Staff working for General Colin Powell, the chairman of the JCS, Kelly would conduct the daily televised briefing during Operation Desert Storm. He became a familiar face for many Americans who found the jocular but direct general refreshing. In 1970, however, he was several months into command of a tank battalion and determined to make it the best in the division. As with many rising armor and cavalry lieutenant colonels, Kelly also had "flash and dash" in his demeanor, a characteristic entrenched in the branch. Driving a bright yellow convertible Impala Super Sport, he cut a memorable figure.

After settling the family with my mother in Kansas City, I returned to the battalion Monday morning as directed. I was more than surprised when Lieutenant Colonel Kelly looked up and said, "Where the hell were you Saturday? I fired Charlie Company commander, and you could have had that job."

Having been told to return on Monday, I fumbled for an explanation before realizing Lieutenant Colonel Kelly was wryly smiling. Kelly said, "If you can manage to find your way down to Charlie Company, tell the First Sergeant that you are the new company commander." It wasn't the traditional change of command ceremony, where the battalion commander would pass the unit guidon from the outgoing commander to the incoming, signifying the passage of command and responsibility; there were not the normal short speeches by the lieutenant colonel and two captains. Rather, Sergeant First Class Wesley Durham saw me enter the office, stood, and said "Sir, I've been expecting you."

Though it wasn't the usual way things were done in the Army, I took it in stride, and inwardly admired the battalion commander, who was certainly a decisive man of action that cut right through any useless bureaucracy. Kelly also seemed to have a wicked but endearing sense of humor. It was about that time I realized that the major back at Personnel Command and Lieutenant Colonel Kelly had probably agreed on the assignment earlier, and there was no way the major would have changed it. Kelly was a battalion commander, and thus a rising star. The major certainly didn't want to get afoul of someone who would probably become prominent in the branch. Looking back, I learned something that would remain with me: Just do the job that you are told; things tend to work out if you do your best.

Wesley Durham was an ambitious and proficient soldier. As a Sergeant First Class, or SFC, he was authorized to be a Platoon Sergeant, like SFC Flores, my platoon sergeant at Fort Hood. However, in the draftee, pre-All Volunteer Force of 1970, some soldiers were a disciplinary handful, others were simply not cut out for the rigorous training and physical challenges, and most wanted to get to the end of their obligation to Uncle Sam with the least hassle and expenditure of effort possible. Those who had been to Vietnam figured they had done enough already; those waiting to go sometimes didn't see the need to train hard on what they thought was "pretend war." This made the first sergeant job difficult, and a few master sergeants, from which first sergeants are drawn, declined or otherwise ducked the assignment. Being motivated and ambitious, SFC Durham volunteered to serve in a position above his rank and was assigned as the Top Soldier of Charlie Company. Durham was a stellar first sergeant, and he and I meshed perfectly. Again, here was a superb NCO who understood that one of his unstated jobs was to advise, coach, mentor, train, and assist his captain to be the best possible. Throughout my career I would tell NCOs that I received my degree and commission from West Point, my training at Fort Knox and in the field, and any savvy from good NCOs, including my father-in-law, the Master Blaster. The casual but professional manner that tended to be my style maintained the professional distance an officer must have with subordinates but interjected the camaraderie and personalism that hopefully caused the NCOs to view me as not only their commander but a fellow warrior who valued them and their work.

In the months that followed, training was rigorous and rewarding. There were all the activities tank battalions thrive on: lots of maneuvering and practicing tactics in the rolling terrain of central Kansas, and gunnery ranges where all weapons were fired for qualification. Tank crews go through a rigorous and intensive set of ranges that are progressively more difficult and challenging. The capstone event was called Tank Table VIII, Tank Crew Qualification, where the crew had to maneuver around the range and engage different surprise targets that would appear thanks to electric activation. Each crew needed to decide which weapons to use: either one of the machine guns or the main gun, a 105mm high-velocity cannon that fired several types of ammunition designed for different targets. The crew was graded on all aspects of the range: movement; target identification; firing accuracy; and various crew members' duties, such as driving or loading the proper ammo. All was timed to determine if it met the stringent standards. There were three qualifying levels a crew could obtain. The first and lowest level was "Qualified," which meant meeting the minimum standards. The second level was "Superior," denoting a much better than average performance. The third and top level was "Distinguished," recognizing the most proficient crews. The fourth category, in which no one wished to be placed, was "Unqualified." Crews that were found Unqualified and had failed the range—or as soldiers termed as a "bolo"—were disdained. Fortunately, most crews qualified the first time through, although some required a second or third time following additional training during their free time. To go into combat, a crew needed to be qualified in gunnery, along with having other tested skills. Having seen both well trained and poorly trained units in combat, I was adamant that the company be ready to confront and defeat any enemy. I took great interest in training my own tank crew, since a tank company commander was also his individual tank's commander and was expected to be able to fight his tank as well as lead the company.

My peers—and occasionally those of higher rank—would ask how I gained the men's respect, especially the NCOs'. The answer, I said, was simple: respect them, train them, take good care of them, pat them on the back when performing well, and kick them in the pants if not. Captain Greg Camp, one of my former Academy roommates and dear friend, commanded an infantry company not far away from Charlie Company. Greg was an Army Brat and a great fellow. One day Greg mentioned that another company commander in his battalion was having trouble gaining his soldiers' respect and wanted some advice. I did not know the officer personally but knew of him, namely that he desperately wanted his soldiers to put him on a pedestal, as he perceived Captains Camp and Jordan were. The answer, I remarked, was for him to stop trying to climb up on a pedestal on his own. If he truly cared about the welfare of his soldiers, applied tough love when necessary, and trained them and the company to do its job well, they would push him atop that pedestal. Ultimately, though, I did not think that individual would ever achieve his goal because he was too focused on himself, not his team.

In addition to the activity at Fort Riley, the battalion, as part of the division, was scheduled for some very special and exciting training; they were scheduled to deploy to Germany for more than a month. The training included an exercise designed to show our European allies and the Soviets that America had the ability to send major forces to Europe. This was done quickly to counter what was seen as the ever-present threat of a massive Russian attack against the North Atlantic Treaty Organization (NATO) and America's Western European allies.

The exercise was called REFORGER, which stood for Return of Forces to Germany—technically REFORGER II. REFORGER I had occurred the previous spring, but with limited units and troops transported to Germany from the States. Rather, it had allowed a division stationed there to move back to Fort Riley, which had lost its division when the Big Red One went to Vietnam. The unit returning to Fort Riley from Germany after REFORGER I was redesignated as the First Infantry Division, the Big Red One. Some of the soldiers who had served in Vietnam and departed when the division left there also went to Fort Riley and rejoined the newly reestablished division; others had been scattered across the Army as needed. REFORGER II would be the first time the concept of bringing a complete division from the continental U.S. was practiced.

REFORGER II was a grand adventure. It was the first time many of the soldiers, including myself, had been in Europe. Flying to Germany in Air Force cargo planes, all marveled at the sights but didn't have much opportunity to sightsee. Immediately upon arriving, we began to draw the tank-battalion equipment from huge humidity-controlled warehouses. The complete set of equipment for each battalion—tanks, machine guns, tools—had been packed in preservatives and stored for emergency use in case of war and exercises like REFORGER. Having the equipment there before flying the soldiers over was necessary because it simply took too long to transport the thousands of pieces of equipment in a division. Even if sufficient aircraft were available, much of the equipment could not be flown due to size and weight.

Drawing the equipment and preparing it for use involved tremendous work. The window of time designated for completing this task—eighteen hours—replicated that estimated available for preparing to fight and stop the Russian hordes before they overran Western Europe. The battalion would work ceaselessly to accomplish this. Given that this was the first time anyone in the Army had ever drawn an entire division set of equipment from POMCUS, or Prepositioning of Materiel Configured in Unit Sets, not everything went exactly as planned. However, fourteen or fifteen hours into the process, Charlie Company pulled the last of its vehicles, fully loaded, from the warehouse.

The exercise was a true adventure, maneuvering across the actual German countryside, including farms and small villages, rather than a separate military training area. With only the major cities off limits, the exercise had the feel of the maneuvering done in WWII. The soldiers were cautioned not to damage German property, which, given the nature of a fifty-ton tank with tracks larger than a bulldozer, was more easily said than done. Though the Army would compensate German civilians for any damages, German civilians' displeasure with American troops could not be so easily reconciled. Nevertheless, the exercise was successful. Some of the U.S. forces stationed in Europe fought alongside the Big Red One, as did some British Army forces. For this particular exercise, the German Army units were designated the enemy, or OPFOR, Opposing Forces. An exciting incident occurred one night when a column of German Leopard tanks stumbled across my advancing company's front. The large German crosses on the tank turrets reminded me of movies I had seen about the war in Europe. A German column had come out of the dark, driving directly across our front, unbeknownst to them until my tanks opened fire with simulated ammunition. It was supposed to be a Meeting Engagement, where two forces unexpectedly run into each other. The better trained and more tactically proficient of the two usually prevails. With the perfect flank shots, the umpires who accompanied all player units awarded Charlie Company an overwhelming victory in the short but sharp and decisive engagement.

The maneuver continued for several more days, with more engagements against the OPFOR. The British forces—dubbed friendlies as opposed to enemy—were in fact somewhat friendly but did not share the Yanks' concern about damaging the countryside. The British acted as if any damage to roads, farm fields, haystacks, or other German property was simply the cost of doing business. But the care that the average German citizen paid to picking up litter, sweeping streets in front of homes, and even clearing twigs and deadfall from forests surprised and impressed most of the Americans, and the host Germans appreciated the care that U.S. units exercised to avoid causing damage. I would learn in the future that as a result, the German population generally held the Yanks in higher regard than the British, which made living in Germany a much more pleasant experience for the American soldiers and families stationed there. It would be some years, however, before the Jordan family had that opportunity.

The final phase of REFORGER was to account for, clean, repair, and turn in all the equipment that had been drawn upon arrival. One major hurdle in the turn-in process was accounting for each of the hundreds of items drawn. Again, this being the first time it was ever attempted, procedures were unfamiliar and, in some cases, not fully thought out. All the companies in the battalion worked around the clock to meet the deadline. Flights home could not be altered, as the Air Force's capacity was extremely stressed moving the entire division over the course of a few days. Three days into the process, I was able to report a 100 percent turn-in, with all items in proper condition. I didn't make that happen—the soldiers did, and I gave the credit to them. Not one to miss the opportunity to brag on my soldiers, I saw the battalion commander during the dinner meal, and after stating completion of the tasks, I added "If you have any tough missions, just send them to Charlie Company!" In usual fashion,

Lieutenant Colonel Kelly grunted an acknowledgment, his expression somewhere between a smirk and a smile. Little did I know those words would come back to haunt me.

The flight home was uneventful but filled with anticipation. While Germany had been an adventure, we were ready for familiar surroundings. The married men, myself included, looked forward to the homecoming, and the single men looked forward to pleasures of their own. While not nearly the same as returning from a year in Vietnam, the Jordans found the reunion memorable. A while after returning, I compared what I had seen in Germany with Nannette's memories of her family's time there. My father-in-law had returned from his unaccompanied tour in Korea, and the conversations between us often focused on Germany, the people, the field time, the food, and definitely the wine and beer. We were starting to build a bond of combined family and professional camaraderie.

Two months after REFORGER, the battalion was back into its cycle of training and maintenance. I felt fully comfortable, and that I had become one of the battalion's more tactically proficient company commanders. I was authorized ninety-one personnel, including myself and four lieutenants. The tank company was small compared to an infantry company of over 150 men, and I could recognize nearly all by sight and most by name. About one-third of the enlisted soldiers were NCOs—a much higher ratio than in other organizations. I found it a wonderful and enjoyable unit to command. It was then that my boasting to the battalion commander in Germany came back into play.

I received word to report to Kelly at the battalion headquarters. Believing it to be some routine matter, I reported to the headquarters, stopping to see the two majors who served as the executive officer, second in command to the lieutenant colonel, and the operations officer, who did maneuver planning and oversaw the development of training plans. Their wives and Nannette had become fast friends, and weekends often found at least two-of-the-three families socializing, despite the rank difference between myself as a captain, a company-grade officer, and the majors, who were field-grade. This was an important distinction. Only majors, lieutenant colonels, and colonels were field-grade officers. One other distinction was a difference in the dress uniform. Field-grade officers had gold oak leaves or "scrambled eggs" on their dress hat visors. This visual difference denoted the officer's higher status as a member of the "brass." In the business world, they would be junior and mid-grade executives. Other company commanders and their spouses rarely socialized with the majors outside of the battalion-wide affairs. The Jordans quietly took our inclusion as a sign of friendship and professional acceptance.

On that fateful day, I did not receive any advance warning from either of the majors, although I detected they knew more than they said. Upon reporting to the battalion commander, I was reminded of my words in Germany. LTC Kelly grinned and said, "You mentioned difficult tasks. I have one for you. I plan to take you out of command of Company C, and give you command of HHC." Shocked, I could only nod. HHC, Headquarters and Headquarters Company, was dreaded by most captains, and I was no exception. I loved Charlie Company and its ninety-one men. HHC was a 312-man monster of an organization. The entire battalion was about 585 men; I would be responsible for over half of those soldiers. The HHC had 292 enlisted, eighteen officers, and two warrant officers. The problem was that most actually worked for someone else directly. Most of the officers—including the lieutenant colonel and the majors—were on the staff, and most of the enlisted soldiers had immediate supervisors and performed their duties outside the company. The enlisted soldiers belonged to organizations such as the dining facility, with its crew of thirty; the maintenance section, with nearly fifty mechanics; or the medical platoon, which worked for the battalion surgeon, usually a captain and physician. There were also other specialty units like the heavy mortar platoon, the air defense platoon, and the scout platoon, similar to the recon platoon I had led in Vietnam. Each had its own officer leader, and each reported to the operations officer, whose responsibility was to train them. Unlike Charlie Company, where I designed and directed the training, I was merely administratively responsible for these soldiers. I was left with the less satisfying tasks: discipline, military justice, and the never-ending administration. While not nearly as gratifying as running a tank company across the countryside, I slowly took pride in accomplishing tasks that had taken no effort in the ninety-one-man tank company. I also learned that if HHC didn't function properly, the battalion wouldn't either: fuel and ammunition would not be delivered; broken tanks and equipment would not get repaired; injured soldiers would not receive care; the battalion command post in the field would not be set up and protected. What I really learned was what it took to keep a fighting force supplied, fed, cared for, and prepared. It took a while for me to realize these things and appreciate what I was learning.

Thoughtful battalion commanders filled the HHC slot with captains who had already proven their leadership and command ability in a "line" or combat company. The battalion commander who relegated the HHC to a captain with questionable ability would learn a painful lesson. The captain would struggle, the battalion would suffer from inadequate internal support, and the battalion commander's stock would sink with his superiors. This lesson was not lost on me, nor were the lessons about large and complex organizations. Belatedly, I realized the faith and trust that LTC Kelly had placed in me.

Meanwhile, my family enjoyed the flurry of activity associated with what the officers considered the "battalion family." Social life revolved around the battalion. For spouses, not only were there the occasional formal dinners and balls but also coffee clubs and afternoon teas. The Officers' Wives Club, the chapel group, and the like were well attended. Monthly hail and farewell events were held at the Officers' Club, or O Club, to welcome and bid farewell to those just arriving and those departing, and both the officers and their ladies enjoyed the camaraderie.

Many battalions conducted another routine activity: weekly "beer calls" for the officers after work. In 1st battalion 63rd Armor, Wednesday at 1800 hours, or 6:00 p.m., unless on leave or in field training, officers were expected to be in the O Club—this was a stag affair. On Fridays, the same applied, except spouses were invited. By that time, the Army had dropped the Saturday half-day of work, realizing that little was actually accomplished to justify it. As a result, the Friday affairs tended to run longer and later. Because many battalions in the division held similar beer calls, the duty rotated between companies to secure sufficient tables in the O Club. Woe be to the company commander who was unable to obtain the tables for the battalion. Normally, a lieutenant was dispatched to seize and hold the critical terrain until the battalion's officers arrived in force. Having eighteen officers in HHC, I usually dispatched a force of at least two lieutenants with orders to "seize the key terrain or die trying." These beer calls, including the co-ed ones, were usually lively affairs; on one occasion, the Jordans joined the live band on stage playing for the crowd. Great friends were made at Fort Riley.

During that time, changes also occurred in the extended family. My father-in-law's parents were getting on in age, and he was trying to decide whether to retire to be near and assist them in Lexington, Kentucky, while being considered for promotion to Sergeant Major, the highest NCO grade. The Master Blaster First Sergeant Owens had all the credentials and an excellent service record. I considered him a sure bet to be promoted.

Ultimately, concern for his aging parents caused Dallas Owens to make the hard decision to retire. In the end, family must come first. While I was saddened by my father-in-law's leaving the service, I understood because of the family lessons learned so many years earlier.

The first months of 1972 saw more of the change and moving so common for military families. Having successfully commanded two companies, I was alerted that I would attend the Captains Advance Course, also known as the Career Course, since officers who didn't resign at the end of their service obligation were committing to more years of service and the possibility of a full military career. The Jordan family repeated the now familiar pattern of packing up, relocating, establishing new friends, and looking to reunite with old ones. As an Army Brat, Nannette was well versed in this; I was growing accustomed as well.

Looking back, I realized that going to Fort Riley after Vietnam had been the best option by far. The lesson learned was that it is the people that make an assignment or place enjoyable or otherwise. I also began to realize that the Army would be my career. While I was still on a route of surprises, I had come to the decision that the Army was my true calling and family business. I just did not yet know how much of my family had or would in the future be in that business.

CHAPTER 7

The Marines, the Big Army, and More

When the Army issues orders, they're not always the final word—assignments can change. I received a call from my branch assignment officer in February 1972 informing me that I'd be attending the Captains Advanced Course and that written orders would follow. It looked like I was headed back to Fort Knox—the "Home of Armor and Cavalry"—where the Armor Center and School was located. But when the written orders arrived, I first thought there'd been an error. They instructed me to report to Fort Benning for the Infantry Captains Advanced Course; a quick call to the assignment officer clarified the situation. A few select Armor captains who had already demonstrated tactical and leadership proficiency were being sent to other branch schools for professional "broadening." I would later learn there was another reason: The Army also used these cross-assignments to showcase top-performing Armor officers to impress other branches with the high caliber of Armor leadership. It was a subtle form of one-upmanship.

Having served in an Infantry position in Vietnam and wearing the Combat Infantry Badge, I felt that my upcoming assignment to Fort Benning would be pleasant enough. Rather than being seen as just another tanker taking up space, I believed my fellow students and the faculty would be more likely to accept my bona fides. I could also learn something that I might not otherwise come across at Fort Knox. I was already familiar with Fort Benning, having completed both Ranger and Airborne training there, and I was familiar with the post and the Army town of Columbus, Georgia. Nannette had accompanied me during Jump School and knew a bit about the area also.

What had seemed like a smooth transition was turned on its head when revised orders arrived. The new instructions directed me to report to Quantico, Virginia, a place I had never heard of. I didn't know what military school was located there. Was it a logistics school? Why on earth was I being sent? Another call to the assignment office brought a surprising answer: I wasn't going to an Army school at all, but to one run by the U.S. Marine Corps. I had been assigned to attend the Amphibious Warfare School (AWS) for six months.

Located about thirty-five miles south of Washington, DC, along the Potomac River, the Marine base was named after the small town it surrounded on three sides—with the river forming the fourth boundary. The town of Quantico was something of an oddity. With a population of about 500 people, it felt more like a relic from the late nineteenth century than the typical "Army town" that usually developed outside Army posts.

When the Marine base was established and expanded around the town, some residents elected not to sell their property to the government and instead remain on family homesteads. The main civilian support community for the base—stores, car dealerships, pawn shops, and other businesses were found in larger nearby towns, such as Triangle, Dumfries, Woodbridge, and Stafford. Those towns also offered rental properties for military families who couldn't secure accommodations on base.

This next adventure for the Jordan's promised to be different. Not only would I be learning dramatically different material, but I'd also be navigating the customs, traditions, and professional culture of the Marine Corps, which differed from the Army.

Atop that was the setting itself. Though only thirty-five miles south of DC, we were still in Northern Virginia—a region with its own distinct feel and atmosphere. Many residents worked in DC, and most were associated with the federal government. The population was largely composed of transplants—people who had moved there for their work. Overall, it was a highly educated community, with a greater percentage of college graduates than the national average. It was also more homogeneously middleclass, though in Northern Virginia that meant something different, given the high cost of living. While not everyone was affluent, there was little of the crushing poverty seen elsewhere. The area also felt fast paced and cosmopolitan. Local news on television was often dominated by stories that would be considered national or international elsewhere. We found a nice but small townhouse about five miles from the base. The neighbors were cordial but not overly engaging, unlike the atmosphere in on-base housing. In the civilian community, most people had their own circle of work friends and social connections, and little effort was spent on trying to meet and get to know the military families who were seen as "just passing through."

The AWS class included about one hundred officers. All were Marines, except for four of us "Army Doggies," as we were called, and one Navy officer—a "Swabbie." There were no Air Force officers in the course. When asked why, the offhand response was, "We don't let them in here." The truth was that the Marine Corps saw little need for reciprocal training or coordination with the Air Force. The Marines had their own Air Wing to provide close air support and some transportation. What it could not provide, the Navy fulfilled from its carrier-based or land-based aircraft, and sea transport. They grudgingly conceded that the Army handled much of the ground combat—but only after the Marines had stormed the beach and secured a foothold. When someone raised the possibility of a land war to stop the Red Army in Europe, the reaction was often blank stares, as though the question had been posed in a foreign language. If it didn't involve a beach or an island to storm, it didn't seem to exist in their worldview.

Naturally, the four Army officers in the class gravitated toward each other. One of officers was Jim Madora, a classmate from West Point. We hadn't known each other well at the Academy, but we grew closer during the course. In later years, we would cross paths occasionally. We would both rise to the rank of general, and eventually reconnect in retirement—partnering on business ventures and professional pursuits. At the time, however, none of it had even crossed our minds.

Both the Marines and the Army officers were focused on the next *real* assignment following the course. AWS wasn't overly demanding academically and given the Marine Corps' strong emphasis on fitness, there was plenty of time to work out and stay in shape. To my surprise, course grades were not displayed. However, posted on the wall was a list of our individual weekly pushup, chin-up, and mile record. The board listing each student's performance metrics was a clear reminder of what the Marine Corps valued most.

Social life during military schools like AWS or the Armor Advance Course wasn't as active as it had been in our previous units, but there were still occasions to meet and mingle. In addition to a few large, class-wide functions, there were informal gatherings held within the smaller sections of about twelve officers into which the class was divided. Spouses were included in many of these gatherings. Naturally, the four Army couples occasionally got together as well. We came from different branches and hadn't known each other previously, except for and Jim and me, who had attended West Point together.

Jim was Air Defense; the other two came from Infantry and Military Intelligence. As it turned out, neither Jim nor I would ever encounter them again after AWS.

Army spouses at AWS had fewer opportunities for unit-based activities than they had in past assignments and, to a degree found some cultural differences among the Marine spouses. The Marine wives often assumed that everyone shared similar experiences, but that wasn't always the case. Army wives were notoriously tough, self-reliant, and resilient. While the Marine spouses certainly shared those qualities, they also projected a more "rough and ready" persona—mirroring the grit of their Marine husbands. They didn't seem to quite share the traditional expectations of what an "officer's lady" was supposed to be.

Another difference was in how each service seemed to view families. The Army had long moved past the old saying that "If the Army wanted you to have a wife, it would have issued you one." With a far higher percentage of married soldiers, the Army had developed programs to address family needs. Some Marine spouses, by contrast, thought the Service viewed families as just another complication. They were also a bit envious of the Army's broader range of duty stations where families could accompany their service member—places like Europe, Italy, Japan, and many posts across the U.S.

While Army families dealt with field time, training exercises, and deployments to Vietnam or Korea, Marines and Navy families routinely faced long separations as their spouses deployed for three- to six-month sea cruises to places like the Caribbean, Mediterranean, or South Pacific. These might have sounded exotic, but they were anything but pleasure cruises—they were filled with intense training.

Another point of envy was how the Army seemed to give families more notice about where and when they would move for upcoming reassignments. That wasn't always the case, as we could have explained—citing our own recent adventure with changing orders.

While the six months with the Marine Corps had been interesting and memorable, my fellow Army officers and I were looking forward to returning to Army life. The Northern Virginia area, and Washington, DC, had been enlightening, but few of us considered it as truly a representation of "the Army." To us young captains, the Pentagon was mostly an abstract or hypothetical place—just a pseudonym for the broader Department of Defense leadership. That view would change in a few years, but for now, we were focused on learning where we were each headed next. This had been a brief stop along my longer journey.

▼▼▼

Once again, I received orders to report to Fort Benning, Georgia, and I wondered why that place seemed so popular with my assignment officer. I wasn't headed to an Infantry unit or a faculty role at the Infantry School, which had been a possibility. Instead, I was assigned to a special staff position. A couple years earlier, the Army had commissioned an extensive study of its training needs, led by a retired general known for his success as a trainer during World War II and the Korean Conflict. In the aftermath of the Vietnam drawdown, the U.S. Army found itself lagging behind its Russian counterpart in terms of modern equipment, doctrine, and training. This prompted large-scale modernization efforts.

One outcome of the training study was the creation of the Combat Arms Training Board, charged with tackling training deficiencies. The Board was given money, personnel, and the authority to develop new training techniques, doctrine, materials, and devices. It was tasked with what would later be called "out-of-the-box thinking." I was assigned to the Armor Team. The year I spent there was intellectually and professionally stimulating. The Board developed many training concepts, requirements, and devices that paid huge dividends for years to come. Some ideas were ahead of their time, waiting for technology to catch up—for example, the use of lasers to safely simulate combat engagements in training. This assignment gave me my first glimpse into how higher-level organizations worked and "Big Army"—the Department of the Army—made decisions and pursued complex issues. These were processes you did not see at the company, battalion, or even division level.

The Board reported directly to the four-star general who commanded all Army units in the continental United States. It also had the attention of the Office of the Vice Chief of Staff of the Army. With that amount of horsepower, the colonel serving as President of the Board operated with wide latitude—and the organization did not hurt for any resources at all.

That time made me feel more intellectually challenged than any time since West Point. The assignment expanded my professional thinking in ways I hadn't previously experienced. My practical knowledge of training and tactical operations helped immensely, and my understanding of what went into major Army decisions grew exponentially.

Later, the four-star command that had overseen the Board—then called Continental Army Command (CONARC)—was in two. One half became Forces Command (FORSCOM), which managed all operational forces within the lower forty-eight states, such as those at Fort Hood and Fort Riley. The other half became Training and Doctrine Command (TRADOC), which oversaw the Army's training schools, developed doctrine, and generated requirements for new equipment. I didn't realize it at the time, but my experience during TRADOC's formation was only the beginning of my association with that organization. I would serve with TRADOC twice more in the future—but such a path was beyond my imagination then.

Our family enjoyed our time at Fort Benning. We lived in one of the newest housing units built for company-grade officers. Like many Army posts, Fort Benning still had many World War II-era buildings that had been intended as temporary wartime structures. After the war, however, the greatly expanded peacetime Army found it necessary to keep many of them in service. Family housing was no exception.

When I reported to Fort Benning, I visited the Housing Office to obtain quarters. A disinterested civilian worker glanced up, saw that I was a captain, handed me three sets of keys, and told me to choose one. It was assumed I was another captain reporting for the Captains Advanced Course and just needed a place to stow my family for a few months. After touring the options, I was extremely disappointed. The quarters were small, outdated, and barely adequate—unsuitable for someone who would be stationed there for a couple years or more. I returned the keys and told the clerk that none of the options were acceptable. Before he could respond, I once more produced my orders and pointed out that I wasn't a student but part of the permanent party, assigned to the Combat Arms Training Board. Whether it is was my unyielding manner or the mention of the Board, I'm not sure—but the clerk quickly fished another set of keys from a different drawer and quietly suggested I check out one more unit. That set of quarters was newly built, had never been occupied, and was perfect.

The family quickly integrated into the new neighborhood. The cul-de-sac was filled with other captains and families, with most around our age, and there were plenty of kids Larry Jr.'s age. Actually, he was known as Larry Reginald, to distinguish him from his father. "Junior" was never a term favored by his parents or him. Life on the large post was pleasant, and we enjoyed our time there as a family.

Nannette became active in several post activities, including the Officers Wives Club and volunteering with the Red Cross. One of her growing interests was developing her public speaking skills. She'd heard of an organization called Toastmistress, which gave women a chance to meet, give prepared talks, and socialize. Though based in Columbus, the group included many military spouses. After some phone calls and written inquiries, Nannette received an invitation to attend the next meeting. They would certainly like to include the wife of a newly assigned captain to Fort Benning.

Before the scheduled meeting, she received welcome call to confirm the location and time—held at a local country club. During the conversation, the caller casually mentioned that the club had a "select membership." When Nannette explained that she was African American, there was a pregnant pause. She was then told that the meeting would need to be moved and that she would be contacted with an updated location and time. The Toastmistress club itself was progressive and open, however the country club obviously was not.

Nannette did attend; she both enjoyed and benefited from her time with the group. She was able to build her confidence as a public speaker. Seeing some of the idiosyncrasies of the "Southern Belles" was an eye-opening experience for her—and hopefully for them as well, encountering an educated, eloquent, and confident Black woman. Later, when Toastmasters International opened its membership to women, Nannette joined that larger organization at future duty stations.

This was 1972, and the many establishments in the South still practiced exclusion and discrimination—despite assumptions that such behaviors were a thing of the past. It reminded us that bigotry and small-mindedness were still alive and, in some places, accepted. There were also those who remained oblivious—not out of malice, but simply because the exclusion wasn't aimed at them. That experience reinforced something I had long believed: You can never let your guard down.

After a short but fulfilling year at Fort Benning, it was time to move again. Professionally, I had completed most of what was required at my current rank. With the drawdown in Vietnam nearing its end, promotions across the Army had slowed significantly. Gone were the days of making first lieutenant in one year, and captain a year after that. To normalize the system, my year group would need to spend about ten years as captains—plenty of time to practice at that rank. Fortunately, I had already commanded two companies, completed joint schooling with another service, attended every required Army course (and more), and gained valuable experience on a higher-level staff.

Still, there were a host of assignments available—many of them necessary but uninspiring. I considered some of them soul-draining. What saved me was a significant opportunity: I had been selected to return to West Point as an instructor in the History Department. Not only would that be an enjoyable assignment, but it also meant the Army would send me to a civilian graduate school for a Master of Arts degree. A fully funded advanced degree and a great follow-on assignment—it didn't get much better than that! All I had to do now was gain admission to one of the Army-approved universities and programs.

Compared to the challenges of applying to West Point ten years earlier, this process felt like a piece of cake. My West Point class standing, grade point average, and Graduate Record Exam scores gave me a solid foundation—and a choice among a number of respected universities. In the end, I elected to attend Indiana University, located in the city of Bloomington.

Before we knew it, it was time to move again. It seemed that we were always either packing or unpacking boxes. This time, as always, the excitement of the next adventure helped outweigh the frustrations of relocating. The move proceeded normally, except for some scheduling problems and slovenly work by the packing crew. I filed an official complaint requesting that the nationwide moving company never handle another one of our moves. True to its word, the Army never assigned the job to that company again.

However, the adventure didn't end there. While driving to Indiana, we passed one of that company's trucks on the side of the road—on fire.

Was that our truck? Were all our earthly belongings the victim of fire and smoke? There was no way to know. Fortunately, our household goods arrived on time and in the same condition we last saw them. We decided that, sometimes, even this world's material things benefit from the power of prayer.

CHAPTER 8

Graduate School: Another World

The Department of History at Indiana University was superb, with several renowned professors—two of whom specialized in U.S. diplomatic history, my area of interest. The 30,000-student campus was beautiful and bustling with activity. We settled into a townhouse in a new Bloomington development whose residents were mostly middle-class professionals. My captain's pay and allowances provided a comfortable—but certainly not palatial—lifestyle.

That said, our earlier housing search had been disappointing and disheartening. We had considered buying rather than renting for two years and were encouraged to find a number of appropriate homes priced around $25,000—a large figure, though modest compared to future housing costs. Confident in our financial standing, being mostly debt free with a guaranteed Army salary, we worked with a real estate agent and were quickly pre-approved for financing.

That changed once the bank loan officer discovered the applicants were an African American family. The loan was suddenly denied on the grounds that "the customer did not meet our loan criteria." The realtor was clearly distressed and apologetic. Despite what was probably a case of informal redlining from local lending organizations, it became obvious the realtor hoped to push the sale through.

Racial prejudice and unequal treatment still existed in civilian life—regardless of the more progressive changes happening elsewhere, and despite the Army's formal stance on racial equality. While I had no illusions that personal bias was entirely absent from the military, open discrimination was not tolerated. I began to more fully appreciate the differences between the military community I had known and the civilian one I had just re-entered.

If bigotry or discrimination existed on the Indiana University campus, commonly known as IU, I didn't notice it. The environment was characteristically liberal, and with a student population that large, diversity was part of the fabric of daily life. Youthful activism served as a natural counterbalance to overtly prejudicial behavior.

Like many graduate students, I was far too busy for the pursuits of undergraduates. But after years of hard work and long hours in the Army, returning to academic life wasn't overwhelming. A stable home life, free of the distractions common in college life, helped me maintain a healthy balance between study and family. At nearly twenty-eight years old, I was older and definitely more mature than many of my classmates. Unlike most of them, I also had a steady income: the Army continued to pay my full salary, and I had a faculty position at West Point waiting once I completed my degree. Most graduate students had neither. In short, life as a sponsored Army officer in graduate school was very good.

A few other Army officers were also attending Indiana University for the same reason I was: They were preparing to teach at West Point. Only one other officer was assigned to the Department of History; another was headed to the English Department, while several were going to the Department of Physical Education. Those in physical education were granted one year to complete their advanced degrees in human physiology. In contrast, English, history, and other academic departments were given a full two years.

This generous timeline provided more than enough opportunity to complete the Master of Arts coursework. In fact, many of us were able to fulfill the credit requirements for a Ph.D., even though we didn't have the time to prepare and defend a dissertation. For those interested in pursuing a doctorate later, this arrangement offered a substantial head start. It was a win-win: Officers received a rigorous academic preparation, and the academic departments at West Point benefitted from instructors with broader and deeper scholarly backgrounds.

The first year passed quickly, and we continued to enjoy the experience. We met several interesting people, though we grew close to only a few. My studies consumed most of my time, but we made a point to visit Nannette's parents in Lexington, Kentucky, just under a three-hour drive from Bloomington, whenever we could. Many weekends were spent there, much to the delight of Nannette's parents, who especially enjoyed seeing Larry Reginald.

His unique status as first—and only—grandchild was about to change, however. Nannette was expecting our second child early the next year. We were all excited, especially the soon-to-be big brother, who was fascinated until he began to wonder if the baby "will want to play with my toys?"

Having a baby on the way revived an old question: What to name the child? The first matter, of course, was the baby's gender. At the time, ultrasound technology was not as advanced as it would later become, so delivery-room surprises were still common.

Nannette had already weighed in on the matter, remarking that she had grown up with four brothers and understood boys. She wasn't as sanguine about raising little girls, who, she said, could be "a handful." I suspected she must be referring to herself and her cousin Charlotte, who was about her age and a favorite childhood running mate. From my own experience with a younger sister, I had always thought of girls as darlings, sometimes a pest, but rarely troublesome. Nannette seemed to view it differently, and spoke of moodiness, "little drama queens," and hormones.

Larry Reginald was excited about having a sibling and especially enthusiastic about being told he could help name his future sister or brother. In the end, fate favored Nannette with a second son. Naming him—Karl Anthony Jordan—was surprisingly easy, and even Larry Reginald approved, although he had originally expressed a preference for a sister. Within a few years, the brothers were inseparable, and Larry was thankful for the way things had turned out.

The second year at Indiana brought new friendships into the Jordan circle. Two African American captains and their families, Captains Larry Ellis and Ishmon Burks, arrived for the one-year physical education program. Upon learning they had arrived, we hosted an informal gathering at our home to welcome them and others.

Among the guests was Captain Robert Dance and his wife. He was pursuing a graduate degree in English and would be joining the English Department at West Point. "Bob," as he preferred to be called, had a particularly interesting background. He had originally been slated to become the eleventh Black cadet in my West Point class. A former enlisted soldier, he had earned a presidential appointment to the Academy. But weeks before reporting, his orders were unexpectedly—and inexplicably—changed, and he was instead assigned to the Philippines Military Academy as an exchange student. He spent four years there and graduated in 1968, ultimately commissioning as an Armor officer in the U.S. Army.

Bob and I became good friends, Years later, I nominated him to be designated an honorary member of the West Point Class of 1968—a nomination that was accepted.

Many of those present at that gathering would become lifelong friends. The Jordan and Ellis families, in particular, had no idea just how close their friendship would grow. We would go on to serve together repeatedly and eventually became part of each other's extended family. The Ellis family had two daughters close in age to Larry Reginald and Karl, and the girls soon became the sisters our boys never had. The Burks and Dances also became part of this larger circle—lives that would remain intertwined at West Point and well beyond.

As is often the case in family life, happy times are sometimes intermixed with sad ones. The autumn of 1974 saw my sister, Joyce, marry her Northwestern University classmate and longtime boyfriend, Jimmie. Larry Reginald was asked to serve as ring bearer, which worried me due to my namesake's boundless energy and knack for distraction, neither of which promised wedding-day decorum. Fortunately, all went smoothly, and Joyce and Jimmie became Mr. and Mrs. Jimmie Moore without any ring-related drama.

Both Joyce and Jimmie pursued degrees in education and became teachers. I jokingly commented that they were going into the family business, following in the footsteps of our mother and grandmother. Then I added that I was too—after all, I was on track to become an instructor at West Point. But by then, I understood clearly that the Army was the true family business of my side of the line.

Some months later, grief enveloped our family with the passing of my uncle, Herbert Johnson. Uncle Herb had always been easy-going, slightly cavalier, but lovable presence in the family—someone who would never let life get him down or interfere with his devil-may-care spirit. He enjoyed a good bottle of whiskey, female companionship, and the art of storytelling.

Unlike my father, who rarely spoke of his time in the Army, Uncle Herb was open about his World War service in Italy. He had served in an anti-aircraft unit, operating heavy guns and searchlights, and deploying barrage balloons—those airborne deterrents suspended by cables to protect harbors, anchored ships, and military installations from aerial attack. On a cold but clear late-winter day, the old soldier was laid to rest at the Fort Leavenworth cemetery. A lone bugler played the bittersweet notes of "Taps," which echoed back. My mother, Dorothy, wept quietly as her brother's casket was lowered. I stood at attention and rendered a final salute to Tech Sergeant Herbert Reginald Johnson.

I also thought of my father, Tech Sergeant Napoleon Jordan. It had been over fifteen years since he and my mother separated, and just as long since I had last seen him. It would be many more years before I discovered why—unlike my uncle—he had rarely spoken of his military service during World War II.

At the time of that war, the Army was still segregated. Most African American soldiers were assigned to support roles in logistics and services, though the Black combat units that did engage performed with distinction. While Uncle Herb had served in an anti-aircraft unit that saw action in Italy, my father had been assigned to a service support unit in the Philippines.

Among his duties was one of the war's most gruesome and emotionally taxing assignments: the necessary function of graves registration. His was responsible for the handling and preparation of battlefield fatalities—both identified and unknown—for proper burial. That disturbing experience left a deep and lasting mark on him. The trauma of war isn't always limited to engagement with the enemy in combat.

Two years at Indiana University ended well for us. All of us enjoyed the experience, both academically and personally. Aside from living through one of the area's extremely rare earthquakes, or the unsettling moment when the townhouse next door was consumed by fire, only one other moment stands out in particular: an exchange with my faculty advisor in the Department of History.

The advisor, a well-published scholar, and ardent Cold Warrior, had taken a genuine interest in my academic progress. One day, during a meeting, he asked if I planned to pursue a Ph.D. in the future. He noted that I had already completed the required coursework and probably wouldn't learn much more in the classroom. I would need, however, to go through the hard work of writing and defending a dissertation.

We had developed a close professional relationship, and a casual friendship, as the professor admired professional career officers. I thought for a moment, then replied, "Having Ph.D. behind my name means about as much to me as what gets chiseled on my tombstone." We both had a good laugh. To my surprise, he agreed. Then, a bit wistfully, he added that being a captain in the Army was probably better than being a professor—perhaps a reflection on his own time in the military during World War II.

The only issue with that exchange was when word of my comment made its way to some of the younger, untenured faculty—many for whom the Ph.D. was the first and necessary step to future tenure and career success. In addition, many were younger than me and earning considerably less than a military captain. Despite that, everything smoothed out. Many of the senior faculty appreciated the discipline, maturity, and work ethic of the military students.

I received my degree on Mother's Day, which also happened to be our wedding anniversary. As we packed up and left Bloomington, we looked ahead to another chapter—this time at another institution of higher learning: the United States Military Academy at West Point.

CHAPTER 9

Back to the School on the Hudson

A well-known saying is that "You can never go home again," taken from the Thomas Wolfe novel *You Can't Go Home Again.* However, throughout my military career, there were several times when both the family and I felt as though we were doing just that—returning home.

In 1975, when we arrived at West Point, that feeling was especially strong for me, but not family who had not been there previously. That point of land jutting into the Hudson River had once been my home. Another familiar saying goes, "Home is where the heart is." In some military family homes, it's not uncommon to see a small sign or refrigerator magnet that reads "Home is where the Army sends you." And indeed, of the twenty-three addresses we would eventually have, West Point became one more place that was home.

Our arrival at West Point differed from any we had yet experienced. Most Army family moves occur in the summer months, when school is out, and the Army tried to support this schedule to minimize disruptions for children. As a result, every post sees a number of moves between May and September. But we'd never seen this many people arriving at the same time. Because of the Academy's rotating three-year staff and faculty assignments, and the need to have new instructors on board and oriented before the fall semester, hundreds of families were scheduled to arrive within a tightly coordinated two-week period. The sheer number was impractical to assign housing on a first-arrival basis, as was the usual method.

To manage the chaos, West Point chose a structured process. All vacated housing units were listed by size and location. Arriving officers with families were placed on a rank-ordered list. However, that presented an unusual problem. Since West Point graduates in a given class share the same date of rank—at least through the rank of captain—they had to be ordered by their Academy class standing from years prior. Non-Academy graduates were ranked strictly by date of rank.

On a given day, families assembled in the Academy's largest lecture hall. In order of rank, officers were called to select the remaining available quarters for which they were authorized. However, authorization came with a caveat: some homes were designated to field-grade officers—usually larger, to include entertainment space—while the rest were for company-grade officers, like captains. Even within those categories, housing varied by numbers of bedrooms—mostly three or four, although an occasional two-bedroom apartment was included. The number, age, and gender of children determined the size of the quarters an officer was authorized to select. Though this process was logical, it was often frustrating for families, many of whom were tired from travel and anxious to settle into the best available home.

A similar class-standing process had been used by West Point graduates when selecting their branches prior to commissioning. Having experienced it before, they largely accepted the process. But for officers commissioned through ROTC or Officer Candidate School, the housing selection method seemed novel and strange. For spouses, Army Brats included, the process was baffling. Many had never seen anything like it.

Some likened it to an auction or raffle—except not everyone's bid carried equal weight. What had once been a jovial, joke-filled process for Academy cadets selecting branches now felt like tension-filled contest. For married majors and captains, the process was no longer academic, it was personal. Spouses, often outraged, impatient, and suspicious, were quick to raise questions, "Why did Mary Jane's husband get to choose before you?"

"Why can't you get an extra bedroom like the Smiths?" There were rarely satisfying answers.

Nevertheless, in most cases, families received adequate housing—even if not in the desired style, size, or location. In our case, we felt fortunate. Rather than competing for one of the smaller company-grade houses on West Point proper, we chose a comfortable split-level house on Steward Army Subpost, about twelve miles north. Once a small Air Force base, Stewart Army featured a runway, administrative buildings, and Post Exchange store and commissary for groceries on one side—and a residential area separated by an extensive wooded area on the other. Some families avoided Stewart due to the commute: a thirty-minute drive over the aptly named Storm King Mountain. With an elevation of nearly 1,400 feet, it offered spectacular views of the Hudson River and Valley. As you descended the southern side, a full expanse of West Point came into view—a spectacular panorama that frequently caused tourists to stop at the overlook for photos.

The only problem was that winter weather was treacherous to drive in. Additionally, some officers viewed living on West Point proper, as close as possible to the headquarters and the Superintendent's grand house, as a mark of prestige. Others, however, preferred the separation from work. It wasn't that they didn't love the Army, but the physical distance helped to enhance family time.

The housing area comprised about seventy-five houses, all accessible through a single entrance from the adjoining public road. It was a wonderful place for kids—self-contained, safe, free of through traffic, and well maintained. With an abundance of potential playmates of all ages and a public school less than half a mile away, it offered a family-friendly environment. In contrast to our previous neighborhood in Bloomington, this was a typical military community. All the spouses understood the unique trials of Army life and were quick to offer help to newcomers.

Several families who moved into—or were already living in—the area would become close friends in years to come. Captain Larry Ellis, whom we had met at Indiana University, also moved into Stewart Subpost. By that time, our friendship with the Ellis family was already in full bloom. Larry and his wife, Jean, were graduates of Morgan State University and knew Lew Jones, who had been my best man at Fort Hood.

The Morgan State connection continued to grow. Captain Harry Lumpkin and his wife, Connie, also arrived that summer. Harry was joining the Psychology and Leadership Department. Like the Ellises, the Lumpkins had attended Morgan State and knew both Larry and Lew. In another interesting coincidence, they had arrived at Fort Hood only a few months after we'd left and had renewed their friendship with Lew there.

Over time, many other Morgan State alumni joined our circle of close friends. Eventually, so many were part of our community that people assumed I had also attended Morgan. When asked what class I'd graduated with, I usually responded, "I didn't attend a real college like Morgan—I went to a reform school on the Hudson." That line usually drew reactions ranging from embarrassment to surprise.

Teaching at West Point was both interesting and intellectually stimulating. All cadets admitted to the Academy possessed strong academic ability, and they responded best to instruction that was well-prepared, organized instruction, and delivered with confident military bearing. Each lesson was taught to four separate groups of cadets over a two-day period. The Academy's commitment to small-group instruction—usually with nine to twelve cadets per sections—meant more instructors were required, and the entire cadet class had to be divided into Day One and Day Two schedules. Large lectures, common at many undergraduate institutions, were rare at West Point and served only to supplement the core small-group format.

Teaching two one-hour sections of cadets each day might sound like a light workload. In reality, my fellow instructors and I found our days were filled with lesson preparation, research, quiz grading, daily oral evaluations for each student, and various other departmental duties. When major exams were administered—or when reports and papers were due—reading and grading took considerable time and effort, including some evenings and weekends. Still, the workload was far from the treadmill of activity usual in operational and tactical units. Weekends afforded breathing room—and opportunities. With New York City less than fifty miles south, Broadway shows and other marvelous attractions were now within reach.

Even during the workday, there was always time to recharge for an hour or so going to the gym or attending a lecture in another department. Captain Ellis and I often enjoyed using that time for vigorous games of racquetball, a sport I had learned—and become addicted to—while serving at Fort Benning with the Training Board. We had begun playing each other at IU.

Larry was extremely competitive and, as a member in the Physical Education Department, very athletic. We would go at each other until fatigue set in, neither wishing to lose the most matches. It was always in good humor, but that natural competitiveness would later play out again in our careers.

I was teaching the standard American History Course to Yearlings—Third Classmen, or sophomores. At West Point, the Department of History offered two core options: American History and European History. Every Yearling was required to enroll in one or the other. The department also taught Military Art to all First Classmen, or seniors—a popular course focused on the study of historic battles and military strategies. I featured specially prepared and beautifully illustrated atlases that depicted in detail the maneuvers of figures such as Caesar, Washington, Napoleon, Grant, Lee, Pershing, Eisenhower, and Patton.

Beyond the core curriculum, the department offered a range of elective courses. Officers who had concentrated in specific areas during graduate school were often asked to design and teach these electives. While my graduate work had focused on U.S. diplomacy, I had also studied Urban History, African American History, and the History of Architectural Movements, among other subjects.

In my first year on the faculty, the department head—a colonel holding the congressionally confirmed position of Professor and Head of Department—asked me to design a course in Black History, to begin in my second year. The task involved significant preparation: researching content, building a syllabus, selecting textbooks, identifying films and guest lecturers, and developing writing assignments, exams, and finals.

Uncertain how popular or subscribed such an elective course in Black History would be, I was pleasantly surprised by the response. Enrollment exceeded forty students—close to the course's cap of forty-eight. Also heartening was the demographic mix: about half of the students were not African American. Many of the white cadets expressed genuine interest in learning about the African American experience—an education that would serve them well as future officers in an increasingly diverse Army.

In many respects, the military, especially the Army, seemed to be ahead of broader American society in terms of diversity and equal opportunity. The country had entered the so-called Post Civil Rights Era, with federal legislation codifying many of those hard-won gains of previous decades. At West Point, the timing was especially meaningful. The Academy was preparing to commemorate the centennial of the 1877 graduation of Henry O. Flipper, the first African American graduate. All of these factors seemed to have enhanced interest in the new course.

The course opened many eyes, challenging misinformation, myth, and assumptions held by both Black and white cadets. Several of my colleagues in the American History program began incorporating elements of the course into their own lessons when addressing relevant topics.

In addition to teaching Black History course, I also was involved with one of the many cadet extracurricular clubs—this one called the Contemporary Affairs Seminar, or CAS. Its primary purpose was to provide African American cadets with meaningful cultural experiences and a forum for discussion. Though the majority of participants were African American, a small number of non-Black cadets also took part. Captain Lumpkin co-sponsored CAS with me, and together we became well known among the African American cadets.

Every Black faculty member at West Point was well known, and many acquired humorous reputations and cadet-assigned nicknames—as only such a jocular and cynical group could apply. Captain Ellis, for example, was a boxing instructor considered a tough guy. More than one cadet dubbed him the "Brown Bomber," a joking comparison to the legendary heavyweight champion Joe Louis.

Captain Ellis's wife, Jean, served as cadet hostess, working with a couple of other ladies in the Hostess Office to arrange trips and hotel reservations for dates, chaperone dances, and foster a cadet social life in keeping with proper decorum. The juxtaposition of the two personalities and jobs—Jean organizing formal dances and Larry teaching punches—was both interesting and amusing. Captain Burks, who taught swimming—a skill some Black cadets considered their nemesis, was jokingly called the "Dark Shark."

As for me, I didn't learn my cadet nickname until years later. Apparently, I had been called "the Oracle." At the time, I was uncertain of its origin or significance. I suspected it was typical cadet sarcasm. Years later, a former student, who rose to senior military rank and held a cabinet position after retiring, confided it was meant as praise for what the cadets saw as a refreshing mix of intelligence and common sense.

By the time I returned to West Point as a faculty member in 1975—just seven years after graduating—the Corps of Cadets had undergone a dramatic demographic change. Instead of ten African Americans entering a class of roughly 1,000, nearly one hundred were now matriculating each year. The representation of other minorities, particularly Hispanics and Asians, had also increased.

The most significant demographic change in a century occurred during my second year of teaching: Women were admitted to the Academy for the first time. The Class of 1980 at West Point, Annapolis, and the Air Force Academy included the first cohort of women cadets, marking a shift in military education. Among them at West Point were the first two African American women to graduate—trailblazers in every sense of the word.

The year leading up to the admission of women to the nation's military academies had been filled with public debate, controversy, and arguments in both Congress and the courts. At West Point, I watched with interest as opinions played out among the cadets and faculty alike. Some cadets thought it appropriate for women to enroll, seeing it as a step toward equality—or, more superficially, to improve the dating scene. Others, however, voiced a more traditionalist, even chauvinistic stance, stating that the Academy and warfare should remain exclusively male. That attitude posed some challenges.

I expected that sort of reaction from cadets—most were young men between eighteen and twenty years old, still forming their worldviews. What surprised me was how many faculty members—perhaps a quarter—showed those views. For my part, I thought the move appropriate and good for the Academy. I reasoned their presence would help suppress some of the juvenile behavior always present in the Corps and, more importantly, teach male cadets how to properly respect female peers as equals—something I believed not all of them had yet learned to do. Even more important, it was a necessary step in preparing future officers to lead a military that was growing more diverse.

For the most part, both female and male cadets performed exceptionally well—academically and physically. While there were natural physiological differences, such as average muscle mass, the women held their own across all areas of performance. One of my more memorable incidents came during the summer training at Camp Buckner, the field site on the West Point Reservation where cadets trained between the Plebe and Yearling years. My job was to teach hand-to-hand combat, a course that emphasized physical agility or coordination.

The only problem we encountered was not lack of ability but hesitation. Some male cadets were reluctant to properly take down or disable female opponents, saying it felt like "hurting their little sisters." Understandably, the female cadets took this pejoratively. To address the issue, I would stop the session and call two cadets forward to demonstrate the correct technique. I deliberately chose a proficient female cadet, preferably an athlete, to demonstrate the proper techniques on one of the hesitant males. She would deliver the appropriate take down, with vigorous chops and kicks. I would then compliment her performance, dismiss her, and resume training. The demonstration made its point. Cadets were shown to assess others based on their performance, not gender or any other bias. They also learned that paternalism is a one-way street.

The female cadets appreciated being treated as equals, instead of fragile or unworthy opponents. Their determination and drive didn't surprise me. I had grown up in a family where women were expected to be strong, self-reliant, and confident.

Instructors at West Point served not only as teachers but also as role models and mentors for cadets. Their families also modeled what Army family life could and should look like. During weekend events and sponsored programs, officers and their spouses performed this important role.

African American cadets especially appreciated time and effort that we and other Black couples made to exemplify these expectations. While many faculty couples also provided such examples, there was a unique reassurance for minority cadets in seeing and interacting with mentors who shared their cultural backgrounds and life experiences—affirming their sense of belonging at West Point.

The three years passed quickly—and memorably. We made many new friendships and renewed others. Several volumes could be devoted to the close Jordan friends and families we came to know during that time. Hopefully, someday they will be, as each has their own story worth telling.

One such story is that of Captain Frederick Black. I first met Fred during our junior years, when he visited West Point as an ROTC cadet from Howard University. Fred was an Army Brat and son of Colonel Black, one of a handful of African American officers who achieved that rank in the 1950s and early 1960s. In a military still marked by segregation and slow integration, these few trailblazing Black officers often sought one another, sharing experiences and encouragement.

That legacy carried into the next generation. Among the still relatively few Black cadets at the Academy during my time as a cadet were Tommy Martin and Jim Flower—sons of senior Black officers—both of whom Fred knew through his father. During his visit, Fred made a point of connecting with them, and a number of us made a point of welcoming the visitor from Howard University. The future Captain Black seemed to know everyone in the Army and everything that was going on in it—a trait he would retain for decades and became a source of teasing among friends. One story that should certainly be recorded is that of the Black family.

Fred Black would go on to serve on the faculty at West Point, eventually becoming a permanent professor and retiring as a colonel. His family's legacy extended even further; his brother Gorham also retired as a colonel, and his son Rick graduated from the Academy and returned to teach. In an interesting coincidence, Fred's wife Sylvia came from a prominent family in Durham, North Carolina. Durham was the same city where Nannette had been born, lived for a time, and graduated from high school while her parents relocated during her senior year. The officers and families we grew close to at West Point would go on to play central roles in our future assignments and became part of what we all considered our extended Army family. These included Captain Ralph Tildon, a fellow African American classmate and dear friend from our cadet days.

Ralph never failed to ask about my mother, a reflection of the bond we shared. She had welcomed Ralph and his wife, Florita, during an unexpected visit to Kansas City. That kind of connection was typical, especially among the "Nine" from the Class of 1968. Many of us had enjoyed the hospitality of one another's families and home.

Our time at West Point marked a professional milestone not only for me, but also for Nannette. She had been volunteering with the Red Cross for several years. During our assignment at Fort Benning, that interest led her to take a volunteer training course at the dental clinic. The experience whetted her interest and a growing sense that this might be more than volunteer work.

Motivated, she took additional courses and passed the New York state exam to become a certified dental assistant. Soon after, Nannette began working as a civilian dentist for the elderly in Cornwall-on-Hudson, a small town between Stewart Subpost and West Point. With Larry Reginald in kindergarten and Karl enrolled in the childcare center, Nannette found real satisfaction in the work itself and the daily interactions with the patients.

All too soon, it was time to move on to our next assignment. The final year at West Point had brought meaningful developments. Although promotions across the Army remained slow, both Captain Ellis and I had received early promotions to major. These "Below the Zone" early promotions were awarded two years ahead of the standard timeline. Early promotions were a recognition extended to only about 5 percent of each group, based on performance and assignment history.

In addition to our promotions, we were both selected to attend the next level of professional military education: Staff College. Unlike earlier military schools, this required a competitive selection process. Only about 60 percent of eligible majors were chosen.

Captain Ellis and his wife, Jean, headed to Norfolk, Virginia, where he would attend the Armed Forces Staff College. Nannette and I returned to Kansas, this time for the Army Command and General Staff College at Fort Leavenworth. Harry and Connie Lumpkin were bound for Germany, while Fred and Sylvia Black remained at West Point. Others from our group of friends were likewise reassigned.

Though we were scattered to the winds across assignments and continents, we had no idea just how many rendezvous lay ahead.

Once again, we packed our worldly belongings and prepared for the next move. Along the way, we made stops in Lexington, Kentucky, and Kansas City, Missouri—breaks in the journey that gave us enjoyable family time. The pattern was established: loved ones would come together for precious and memorable moments before scattering to the wind until the next reunion.

Such was Army life. But this time it felt different. As a newly minted field grade officer, I traveled not just as Larry Jordan, but as Major Jordan. I had joined the ranks of the Brass—as a very junior member.

CHAPTER 10

Joining the "Brass"

THE ARRIVAL AT Fort Leavenworth differed greatly from that at West Point three years earlier. With over 1,000 new students arriving to attend the Staff College each year, the Fort Leavenworth Housing Office preemptively assigned quarters based on projected date of arrival, dates of rank, and family size. Having been promoted only two months earlier, I was at the bottom of that list. The Army deemed my two sons—eight and four years old—able to share a bedroom, so they assigned us a tiny two-bedroom house of barely 700 square feet. As fortune would have it, more desirable quarters would occasionally become available due to officers who could not make it to Fort Leavenworth for some reason. A waiting list was established for that purpose. Most families opted to remain in their originally assigned houses rather than move a second time in the ten months the course ran, but we eagerly accepted the chance to move to a three-bedroom house one block away. At slightly over 900 square feet, it wasn't spacious, but certainly better than what we had occupied. The catch was that the Army, which paid moving companies to pack, ship, deliver, and unpack a family's household goods when moving from post to post, would not do so for moves of convenience between quarters on the same installation. Using handcarts, cars, and a borrowed pickup truck, we moved in what would have looked like a gypsy caravan to outsiders. But this was a military community, and a couple of neighbors—majors also attending the Staff College—lent a hand.

The two officers and I became friends and decided to carpool each day to class at the Command and General Staff College at Bell Hall. Often, we walked rather than drove the mile together, as wives would have places to go in the single vehicle most military families had. The walk became a regular thing for me after Nannette took a job as a dental assistant with a civilian dentist in Leavenworth, the town neighboring the post. Years later, those two neighbors and I would all become general officers.

Most couples had children, so the neighborhood seemed to always have two or three packs of kids energetically playing games, climbing trees, or having shouting contests, or racing Big Wheel. A craze at the time, Big Wheels, oversized, low-slung tricycles that could be peddled at amazing speeds, made quite a racket. The interesting thing was that a Big Wheel could accommodate an adult rider, if not too tall or heavy. Some days, we officers would return home to find the wives in the neighborhood riding the Big Wheels in mom races, sometimes fueled by afternoon cocktails or wine. This was just one example of the good-natured relaxation in which we and our families indulged during what the college faculty called "the best year of our lives," referring to the professional development. We students saw it as a great year for different reasons.

The Staff College's purpose was to prepare officers for higher-level planning and staff assignments by focusing on operational principles and strategy. There were lessons on strategy and tactics, exercises in preparing and briefing tactical plans, and both manual and computer games to simulate executing those operations. As with other Army courses, students were assigned to sections, which amounted to permanent class groups. There were twenty-two such sections, each with between forty and fifty individuals. It was a diverse student body of over 1,000 overall, including a mix of American Army officers of all branches, some Air Force, Navy and Marine exchange officers, plus allied officers from several countries.

My section—Section Thirteen—contained the normal assortment of different branches, services, nationalities, and different personalities, such as the studious, the lighthearted, and the always-appreciated practical jokers. The section had several combat arms officers, already versed in tactics and maneuver concepts, along with logisticians, administrative and other specialists, and even doctors and clergy. There were many interesting individuals in Section Thirteen, including a female lieutenant colonel nurse, who was the senior officer in the entire class, and Chaplain Matthew "Matt" Zimmerman, who became a good friend to me and many others in the section.

An outgoing and personable Baptist minister and Army chaplain, Matt Zimmerman never met a person he didn't wish to engage in conversation. One might assume Zimmerman to be a name connected with deep European roots, but Matt was African American with a cheery, homey personality and just a hint of mischief. He offered me a deal. He would put in a good word with The Man Upstairs if I would help him make it through tactics. Years later, Matt Zimmerman would be a two-star major general and Chief of Army Chaplains. I would enjoy teasing him saying that without my help with tactics, he would not have made it past major—and asking him when he was going to make good on putting in a word for me "upstairs."

I had known several members of the class prior to arriving at Fort Leavenworth, and I would serve with them and others I met there in years to come. It was said that the top half of majors went to the Command and General Staff College, and attendees could anticipate serving together as they rose in rank. Some were my West Point classmates, and others I had met at West Point while teaching. Major Al Vitters was a classmate who had been in my cadet company, and we had shared the past three years on the Academy staff and faculty. Al had studied various aspects of female physiology in depth to prepare for integrating women into the Academy's physical training, and I thought him a primary reason that had been successful. Major Rick Rhodes was another classmate, who had attended New Mexico Military Academy before entering West Point, with an easy-going and pleasant heartland personality. I coached his eight-year-old daughter, Kattie, along with Larry Reginald in youth soccer—though at that age, many games were more of a mad scramble than a competitive match. A couple years later, Rick and I would serve again together, our kids playing a better grade of soccer.

The memories made at Fort Leavenworth would remain with my family. For the first time in many years, we were close enough to routinely visit my mother, and the boys got to spend time with their Grandma Dorothy. Some of the memories would make us shiver in the future. This trip to Kansas also brought winter weather reminiscent of time at Fort Riley. Given a bordering Missouri River tributary, the cold wind blowing across the post at Fort Leavenworth had a particular bite. However, unless the windchill was truly arctic, that didn't stop the student officers from pursuing their morning jog. Running was just another part of Army life. At posts like Fort Hood and Fort Riley, commanders led their units on a daily run to start the day, with sergeants calling cadence to keep everyone in step, or singing jingles known as "Jodie calls." While running some mornings I would chuckle thinking about double-timing everywhere as a New Cadet in the summer of 1964. In the sub-freezing temperatures of the Kansas plains, I would put the thought of that hot July out of mind as quickly as I could. The brisk wind also didn't deter me from one detour I took every ten days or so to run to the Fort Leavenworth Cemetery, where I would pay respects to Tech Sergeant Johnson—Uncle Herb—with a few words and a salute. A classmate had seen me saluting one morning and later asked what I had been doing. I replied that I was paying respects to a "good Old Soldier, who had been in the family business."

As Christmas approached, my family's spirits were buoyed by not only the holidays but the possibility of heading to Germany after completing the course. Considering my two trips to Germany during REFORGER II and REFORGER III and Nannette's fond accounts of her family's time in Germany, I thought it would be both professionally rewarding and enjoyable for the family. The only problem was the lack of orders and assurance we would be assigned there. I had an interview with a lieutenant colonel attending a two-week Pre-Command Course segment at Fort Leavenworth. He was headed to the 3rd Armored Division in Germany to command a tank battalion, and, like many of his compatriots, was contacting majors in the Staff College per Armor and other branch Assignment Officers' recommendations who might be potential candidates for executive or operations officer positions in their future units. Many new battalion commanders liked to bring in a few folks that would be on their team from the beginning, rather than solely inheriting the former battalion commander's close subordinates. This informal process was not final, but it certainly helped get the assignment process started.

When the Staff College ended its last session before the holiday break, a few officers were beginning to receive notification that specific orders were on the way, but I had yet to hear anything from the branch. I had arrived home to tell Nannette that I had not received word about assignment yet, but then the phone rang. It was not my branch assignment officer, but another student in the course: Major Tom White, a West Point graduate a year ahead of me, and an armor and cavalry officer. Tom White was known as a mover and shaker, already having a reputation as a fine—though perhaps a bit flamboyant—cavalry officer who seemed to know everyone and everything that was happening in the Army. White merely said, "I thought you might want to know that you are headed to the 3rd battalion 33rd Armor in Kirch-Göns, Germany." I asked how he knew. White just chuckled and said, "Don't ask." That was a typical response, as it only helped deepen the mysticism Tom enjoyed. True to reputation, Tom White's information was accurate.

Major Tom White went on to have an interesting career and was later selected to be a brigadier general. He retired from the Army and pursued a very lucrative business career for some years. Much later, when I was in my thirty-second year of active duty, White would be nominated and confirmed as the Secretary of the Army. At that point, there was more truth to the claim that Tom White knew everyone and everything happening in the Army.

The preparation for the move to Germany differed considerably from the others I had experienced. I had to secure passports for myself, Nannette, and the boys. We knew Larry Reginald was old enough for his own passport but had to research to discover that five-year-old Karl would also need one. That made him feel grown up. There was also the matter of what possessions to take to Europe, and which would remain in commercial storage paid for by the Army. There was a restricted weight allowance, and certain large items—such as furniture and kitchen appliances like refrigerators and ranges—could not be transported. Army quarters, both stateside and overseas, had those items provided—and the upside to this was the provision of well-made German furniture with the quarters there. The electrical current in Europe was also different. American devices would not work on the 220-volt European power, but Americans could purchase portable transformers to convert the European current into something useable for small devices like radios or toasters. Electric clocks, however, would not keep proper time as the electricity was not the standard sixty cycles per second found at home. Some items like that had to be purchased after arriving, but that added to the novelty of the experience.

Shipping automobiles to Europe was another adventure, as they needed to be dropped off at ports either in Bayonne, New Jersey, or Charleston, South Carolina. The Army contracted ships to haul automobiles, household goods, and other materiel for use in Europe. Once families were settled in quarters and their soldiers signed into the new unit, they would receive notification when the vehicle arrived in the north German port of Bremerhaven. The train trip up to northern Germany, followed by the drive home—sometimes for the first time on the infamous *autobahn*, or superhighway, which had no speed limit in most places—was certainly an adventure. However, before being allowed to drive, soldiers and family members had to present a valid stateside license and pass the written German driving exam. This exam was rumored to get the best of many people, requiring multiple retests. Fortunately, most of Europe drove on the righthand side of the roadway like in the U.S. and Canada, the United Kingdom being the exception, but rules of the road and driving regulations differed considerably from those back home. Thanks to diligent preparation, we passed the test the first time.

For the Jordan family, those European adventures were yet to come, with one last adventure remaining on this side of the Atlantic. Following a road trip much like the one before Vietnam, we were again to enjoy brief but memorable time together, stopping to see family before bidding farewell. The Army contracted entire airliners to fly families to Europe during peak summer months. Reporting to McGuire Air Force Base in New Jersey to make our flight, we along with a couple hundred other passengers learned that our DC10 aircraft had been grounded. There had been troubling accidents with the DC10 the past several months, so a switch of aircraft type was welcomed news. However, as the group was bussed to the replacement aircraft, people gasped as they caught sight of the C141 cargo aircraft. These large four-engine jet planes were modern but spartan, used to transport cargo and sometimes paratroopers for jumping. I reassured my wife that everything would be fine, despite the cargo plane cabin lacking the normal airline windows. It would be like being in a windowless tube. I knew the Air Force had removable passenger seats, somewhat less comfortable versions of those found on civilian flights, which could be installed when needed. To everyone's shock, the plane contained only web strap seats, or "jump seats," used to transport paratroopers on short flights to drop zones. Evidently, the demand for replacement planes was so great that not enough passenger seats were available. It was impossible to get comfortable on the web strap seats, which was a complaint that spouses, children, and soldiers alike lent voice to. At the end of the cargo cabin was a boxy contraption known as a comfort pallet, a large port-a-potty on a steel pallet. That and a bag lunch amounted to the inflight amenities for the seven-hour trip to Frankfurt, Germany. It was an inauspicious start to our European adventure.

CHAPTER 11

Germany: The First Trip

The Jordan family arrived at Rhein Main Air Base outside of Frankfurt, Germany, in late summer of 1979. I was accustomed to arriving at a new duty station after serving in the Army for eleven years, but Germany was truly an exciting new world for me. REFORGER had provided a brief snapshot, but now I was to become part of the community and immersed in daily life in a country not my own. Fortunately, I had studied German for two years at the Academy and been required to display reading comprehension in a foreign language for the Master of Arts in History program at IU. Though I'd never understood why an American history major had that requirement, I was now thankful. That background helped immensely. I found that Germans, whether military counterparts or ordinary civilians, appreciated the attempt to use the language, especially if done with any fluency. In most cases it wasn't necessary to speak German, as most Germans received English instruction in elementary and secondary school. Still, the use of the German language was usually received with admiration and, in some cases, greater cooperation or assistance in both important and trivial matters.

The unspoken irritant for many Germans, which I eventually discovered, was that they felt too many Americans acted as if Germany was still an occupied country rather than an ally. While I did not find that to be true, American behavior sometimes lent to that perception. Some Americans simply shut themselves away from the wonderful sights and culture around them. They preferred to remain on the American military bases and housing areas, comfortable in their enclaves. The reality was that those individuals had a small view of the world, or simply lacked the confidence to engage with different cultures, people, and things. It was a wasted opportunity to learn, grow, and experience the richness of the world. I occasionally reflected on growing up in Kansas City and wondered how many of my former classmates and friends were similarly limited in their small worldview. I had not maintained contact with more than one or two, and those had left the hometown at some point. With many family activities on the military installations, it was easy for those who wished to enclave themselves to do so. Still, quite a few Americans enjoyed the opportunities offered to travel and see things only read about or seen in pictures previously. The Jordan family made it a point to learn as much about our host country and the rest of Europe as we could.

Housing for most American Army families in Germany differed greatly from stateside. Instead of single houses, duplexes, or townhouses, most lived in stairwells, large apartment-style buildings, usually three floors high. There were often three separate entrances on the front of the building, each opening into a small hallway with a staircase. On each side of the hallway was a housing unit with three or four bedrooms, an adequate living room, dining area, and kitchen, and usually two bathrooms. Each unit had an outside balcony, a feature nearly ubiquitous in German residential architecture. Many were festooned with flowerpots, as the Germans seemed especially fond of growing lovely plants. The arrangement with two units on each floor meant that six families resided in each stairwell, for a total of eighteen families in the building. In the basements were communal laundry facilities with multiple American washers and dryers, along with storage areas for each family. Husbands and teen boys, who were usually assigned mowing duties, did not miss the lack of personal lawns. While different from which most Americans were accustomed, the stairwells were comfortable. Large common areas behind most of the stairwell buildings provided play space for the children. As in most military communities, kids found plenty of playmates.

Large housing areas often had elementary schools run by the Department of Defense School System, or DODSS. These schools employed American civilian teachers who enjoyed the life in Europe. All children of Americans assigned to Europe could attend these schools, which were accredited like those at home to ensure students did not fall behind academically while abroad. While elementary schools were near most troop concentrations, middle and high schools were located in perhaps ten major cities with a large U.S. presence, sometimes requiring students to be bussed from outlaying installations. These American high schools were rivals in football, basketball, baseball, track, and other sports in the way that cross town schools were back home.

Upon arriving, the family was welcomed into the battalion and brigade in a typical hail and farewell event in the Brigade Executive Officer's spacious yard, furnished with German-style outdoor tables, coverings, and chairs. As usual at such events, food and drink were served along with jovial comradeship. The 1st Brigade units were located at a garrison installation—which in Germany was called a Kaserne, meaning barracks—close to Kirch-Göns, a small village about twenty-five miles north of Frankfurt. Ayres Kaserne—or, as it was nicknamed, "The Rock"—housed five maneuver battalions, three tank and two mechanized infantries, an air defense unit, a maintenance company, and the Brigade headquarters. Ayres Kaserne had the largest single concentration of combat troops in West Germany. The field artillery battalion that supported them was located at Schloss Kaserne in the town of Butzbach, three miles south of The Rock. Butzbach was a beautiful town with half-timbered houses, an old town square with a carved fountain, and a heritage as a Roman outpost centuries earlier. Adjacent to Butzbach was the main American housing area, Roman Way Housing Area. That is where we lived, in a stairwell on Texas Street at the front of the housing area, along with all five maneuver battalion commanders and nearly all of their majors. A few captains were sprinkled along Texas Street, and many more along the streets beyond Texas. Again, new friends and associates were made; we would encounter many of these friends later. One family in particular would follow us for many years to come.

▼▼▼

Captain John Altenburg was an Army lawyer—a member of the Judge Advocate General Corps—welcomed at the same hail and farewell that I was. JAGs, as they were commonly called, had been a part of the Army since the Revolutionary War, and were essential for military justice matters as well as administrative law. JAGs also advised commanders in such topics as international law and the intersection of U.S. and German law regarding behavior by service members and families. Before attending law school, John had been an enlisted soldier in Vietnam. His background enabled him to be an excellent Army lawyer, who understood the operational Army; he also understood the perspective of junior soldiers, who sometimes found themselves afoul of military justice. The Altenburg family was from the Cincinnati area, where many German immigrants had settled in the last century. Some of John's ancestors were from a small village near Dresden; the other half were solidly Irish American. John inherited a taste for both good German beer and Irish spirits. He and I enjoyed a jocular camaraderie, where our different yet complementary personalities blended well. We would have several assignments together, and both families would become close. The Altenburgs were also to have a son and daughter graduate from West Point, and two grandsons attend.

▼▼▼

I arrived in Germany two months before the incumbent executive officer of 3rd Battalion, 33rd Armor Regiment was supposed to depart, so I was assigned to the brigade staff as the S-5, or civil-military staff officer, in the interim. My task was to ensure that public relations between the Brigade and the local civilian officials remained coordinated and cordial. A major part of that involved investigating and arranging compensation on behalf of U.S. Forces for the damages made to Germans' property. Yet, like many things, the compensation system was the butt of some jokes. A common saying among the troops was that if you ran over a German chicken, Uncle Sam had to pay for the chicken, all the eggs it might have ever laid, and the chickens that didn't hatch as a result. The office had two other individuals: a German lady who served as the Administrative Assistant, and a U.S. Army NCO whose mother was German. Both were native German speakers, and most conversation in the office was in that language. The two months spent in that office greatly enhanced my vocabulary.

In every aspect I felt ready for assignment as executive officer and second in command of a tank battalion, where the job was to make certain all aspects of the battalion and its staff were prepared to function at the commander's order. My experience leading three different platoons and two different companies but also teaching and training cadets supported that feeling of readiness. Developing new training doctrine, literature, and devices also helped prepare me for this new leadership role. Plus, I meshed well with the battalion commander, who was the same lieutenant colonel who had conducted the interview at Fort Leavenworth. We both had two sons, the respective boys similar in age and interests. Our wives became good friends and teammates as the senior ladies in the battalion.

Socially, spouses had the usual unit-based activities and Officers' Wives Clubs, plus AWAG, which stood for Americans Working Around the Globe. AWAG was a non-profit organization dating from 1946, when American families were first allowed to accompany their soldier to Europe. AWAG's mission was to connect and strengthen American-affiliated communities around the globe; it sponsored training and other programs for spouses, along with conferences and trips. Spouses had sponsored trips to various European tourist sites in Germany, France, Italy, Belgium, and the Netherlands. Later, after the fall of the Iron Curtain, they also traveled to Hungry, the Czech Republic, and other eastern European countries. Service members had less time for touring but could still see a lot of Europe while on exercises or traveling with family on leave, either on their own or with the many tour bus operators. Apart from organized activities, there was also the opportunity to shop for once-in-a-lifetime treasures such as crystal, china, pottery, art, clothing, and more—from all over Europe. Like many others, we began a collection of mementos. In our case it was antique German clocks, which eventually were to include two grandfather clocks and several wall and mantel clocks.

The move to Germany also reunited us with some of our close friends, the Ellis and Lumpkin families. We were scattered between the Kirch-Göns/Butzbach area, Heidelberg, and Darmstadt. This necessitated weekend drives to gather, but we managed to see each other once or twice a month in one location or another. We also went on some of the bus tours together, which included a few humorous incidents. Once, young Karl would go to use the restroom on a luxury German tour bus, only to not return. Upon unlocking the restroom door to exit, Karl would discover the light went out, leaving him in complete darkness. He instantly relocked the door to turn the light back on but could not figure out how to leave the small compartment without the comfort of seeing what he was doing. After some time had passed, Karl's mother retrieved the confused and shaken youngster.

Another memorable incident occurred when one of the wives did not have her passport while on a bus tour that would cross several borders.

Thoroughly criticized by the group for potentially derailing the tour, her husband told her that they would "pick her up on the way back." This effort to break the tension with humor did not go well for him. Approaching the first checkpoint, she announced a plan of action. She produced her shot record, a small booklet used to record immunizations, proposing to wave it when passports were asked to be shown. Sometimes the border guards would ask everyone on the bus to raise their passport rather than examine each. After all, this was a bus full of American service members and families on a sponsored tour, and these were all NATO allied country borders being crossed. Fortunately, the border agents merely stepped aboard the bus, confirmed the tour was as scheduled, and waved the bus through the checkpoint. The ease of crossing the border did not dampen the group's good-natured ribbing of the relieved spouse, nor did it stop renewed threats of leaving her at the border if more diligent agents were manning the checkpoint on the way home.

In all regards, the year and a half spent at Kirch-Göns in the 3rd Battalion, 33rd Armor was a professionally enjoyable and rewarding success. The usual training events took place, including gunnery at the Grafenwöhr training ranges and maneuver training at Hohenfelds Training Area. Both had been German military training areas since Prussian days. By the mid-point of the three-year overseas tour, there was little not seen, done, or mastered in the operation of a tank battalion. The lessons learned were instrumental in preparing to command one in the future. I met many people, some of which I would encounter later, including the captain commanding the maintenance company at The Rock that supported the five maneuver battalions. Years later, that captain would be the Logistics Officer for all of United States Army Europe (USAREUR) as a brigadier general, and later the senior logistics officer, or deputy chief of staff for logistics, for the entire Army as a three-star general. In both positions, I would remind him that all his work repairing the broken tanks I sent him at Kirch-Göns accounted for his expertise and rise in rank.

After eighteen months, it was time for another major to serve as battalion executive officer, and the family was facing another potential move. There were other positions I could fill at Kirch-Göns, but none was as challenging or prestigious as the one just occupied. Plus, the Army valued former battalion executive officers, considering them fully qualified at their rank level and desirable for higher staff positions. The Armor Branch had notified me of a tentative assignment to the headquarters of the Army's total forces in Europe, the USAREUR, which was commanded by a four-star general. Though the orders had not arrived, I understood the job was mine if I wanted it.

As the time approached to move, I was instructed to meet with the two-star division commander at the headquarters in Frankfurt. That commander was Major General Walt Ulmer, a well-respected armor officer with a reputation as a superb leader and motivator of soldiers. He had been the commandant of West Point Cadets as a one-star general when I was an instructor on the faculty. We had met occasionally, and the general remembered. It was not unusual for General Ulmer to meet with field grade officers about to change jobs—those who had done well were usually invited to meet and discuss future assignments. When Major General Ulmer asked if I would like to join the division staff, which would be a good career move in most circumstances, I informed him that Armor Branch had indicated I was being considered for the USAREUR staff. Ulmer then mentioned he could argue to keep me in the division, if I wanted. Somewhat fearful I might be burning a bridge, I stated that the position as a War Plans Officer for the entire theater at USAREUR might be both educational and beneficial. Surprisingly, Ulmer concurred that I was making the right decision to take the most challenging job, even if it deprived the division of an experienced officer. Two lessons were learned from that meeting. First, always seek the harder job. Second, it is neither disloyal nor inappropriate to offer your recommendation to your boss, even if you think they prefer a different course of action.

The family eagerly anticipated the move to Heidelberg, as it was a wonderful place to live. We had learned of its university, castles, and other attractions while visiting our friends the Ellis family, who were stationed there. Two weeks prior to Christmas of 1980, I made the trip to Heidelberg, leaving the family in Butzbach as quarters were not available at our new assignment until early January. While not as long or remote as other separations, we all looked forward to getting the family under one roof again. When I returned home for a few days over Christmas, Nannette and I planned the move. The move was smooth and well executed, exclusively due to Nannette, who handled it with practice, having grown up in an Army family. She would sometimes say, only half joking, that husbands only showed up for the move; wives did the work and made it happen. I knew that to be true. While I did some manual labor, she was the stalwart that got the family organized, made detailed arrangements, and kept the children from becoming "helpful" pests.

The housing area where we lived in Heidelberg was named Patrick Henry Village, or PHV. It had many large stairwell apartment buildings similar to those in Butzbach, and a couple dozen single family houses designated for the several general officers working in the area and a few of the most senior colonels on the staff or in the various command positions. I sometimes passed those homes on my morning jog before work, but never thought about the possibility of residing in one. Unbeknownst to me, that was to occur several years in the future.

Exciting and educational activities filled the next year and a half. Reporting to the assignment as a War Plans Officer also brought another surprise: my West Point and Leavenworth classmate, Rick Rhodes.

Working in the same office, we prepared and maintained an assortment of theater-level war plans to include the main plan for defending Europe against a Soviet attack. There were other plans to forcibly open a corridor to Berlin if the Soviets attempted to blockade it as they did for nearly a year starting in 1948, and for the use of force by NATO in something less than general war. Numerous high-level exercises were conducted to test the plans, and I learned how a major command coordinated and conducted operations at the strategic level. We were applying the classroom lessons from the Command and General Staff College as close to real warfare as possible without actual combat. It was like a graduate school for warriors.

Some of the most memorable times involved personal trips to Berlin. Lying deep in the sector of Germany originally controlled by the Soviets after World War II, Berlin had been partitioned into four zones, with the United States, the United Kingdom, France, and the Soviet Union each controlling one. The allies eventually combined their sectors into West Berlin, while the Russians occupied East Berlin. The Berlin Wall formed an ugly division between the two. Barbed wire, a concrete wall, guard towers, and menacing East German guards were there to prevent people from escaping to the West, although many attempts, some successful, were made over the years. As the West German government could not claim Berlin as the country's capital, it was therefore in Bonn. The Allies were responsible for defending West Berlin.

In accordance with the Four-Power Agreements reached at the end of the war, special orders had to be issued for American officers and families to visit Berlin. These were called Flag Orders, as they had the four flags of the U.S., Britain, France, and the Soviet Union printed across the top, and they contained precise information about each traveler in the family. The Soviets at the checkpoints would scrupulously inspect these orders—a single period missing or misspelled word was a cause to deny passage. The two ways for families to visit Berlin were by the official U.S. Army

Duty Train or by personal car. By far, the most scenic and interesting was by car, which was also a political lesson for the children.

Passing through the American checkpoint at the border, Checkpoint Alpha, was painless, as the U.S. guards checked to ensure Flag Orders were in hand and that the vehicle had sufficient fuel to make the three-hour trip to Berlin. Once in East Germany, American vehicles were prohibited from leaving the road except at authorized rest areas, and travelers were prohibited from talking with any non-U.S. personnel. Less than one hundred meters across the border was the Soviet checkpoint. Again, orders were checked, but this time with typical communist bureaucracy and attempted intimidation. A Cossack-looking guard demanded the orders. The Jordan family, excluding myself, had to show their passports through the car windows. American military members were not required to show passports, as their uniform was their identification per the Four-Power Agreement. After an authoritative show of inspecting the documents, car, and passengers, I was given the orders and motioned to a small building.

Upon entry was a small, barren room with small portraits of Marx, Lenin, and the current Soviet premier on the wall. There was a single teller's window with an opaque covering and a narrow opening. A heavily accented voice demanded "papers!" Documents handed through the opening were snatched, and a few minutes later came the sound of furious stamping. The documents, heavily stamped with red Cyrillic characters, were shoved back through the opening. I repeated the earlier encounter with the guard outside. He looked at the stamped papers, looked at the car, looked at me, and finally handed the papers back. The Soviet Union had demonstrated its might and authority to control access to Berlin. However, the last thing the guard did after handing the Flag Orders back was attempt to sell me a Red Army belt buckle with a red hammer and sickle as a souvenir. This happened to many American service members. The need and desire for hard Western currency motivated even the most disciplined of Russians to seek a deal with the "decadent capitalists." After my first exposure to the offer caught me off guard, rather than a polite "no," I began responding with "I'll get mine the hard way," meaning on the battlefield. From their reaction, it was obvious the guards understood. They had noted and could tell from the devices and awards on my dress uniform coat and the combat patch on my right shoulder that I had fought in Vietnam. I returned to the car chuckling, and my wife reprimanded me for "messing with the guard," which she worried could have caused us trouble.

The entire process made an impression on the Jordan boys, who saw the guards, the border, and the treatment of German citizens attempting to cross the border in either direction. The Germans had to pass through the East German checkpoint rather than one manned by Russians, and they were frequently harassed or denied entry or exit after waiting in long lines. It was a show of authority and defiance, since the U.S., British, and French militaries did not recognize the East German authorities. The lesson I took from those encounters was that the Red Army was far from the disciplined juggernaut it was often portrayed to be—it behaved differently when no one was watching. Decades in the future, the vaulted Russian Army would reveal its lack of qualities essential for efficient yet ethical operations.

West Berlin was a vibrant and exciting place, especially compared to the drab and utilitarian East Berlin. Still, East Berlin had its own sights and places to visit—though it was also a city of stark dichotomies. In the major shopping areas were trendy shops offering good quality products. What was not immediately obvious was that these shops were open only to Western shoppers and the Party faithful. Ordinary East Berliners shopped at stores off the main avenues and away from the trendy shops. In the controlled economy of East Germany—or, as it was officially known, the German Democratic Republic—all prices were uniform and set by the government. However, the quality of products offered in the shops reserved for the Party bureaucrats and Western tourists was much higher than that of similarly priced items available to the general public. It was part of the communist attempt to convince Westerners that their system provided well for the Proletariat.

During our first tour in Germany, we saw several communist attempts to control all aspects of social and business life firsthand. One example was the money used in East Germany. The currency in West Germany was the Deutsche Mark, or D-Mark, a stable and respected currency that traded on the world exchange. The East Mark was virtually worthless on the world exchange, yet the Soviet and East German governments demanded that it trade on a one-for-one basis with the D-Mark in East Germany. Americans quickly learned to exchange their D-Mark for East Marks before crossing into East Berlin, as they could often receive a dozen or more East Marks for each D-Mark. With the inflated exchange rate, American families felt wealthy when shopping in East Berlin, all the time laughing at the great communist "workers' paradise."

Another example occurred as we were shopping. Upon entering a large department store, each individual or party was handed a plastic basket to hold items selected. The store employee, whose only state-supported job was handing out baskets, did not understand the concept of window shopping, and was perplexed when Westerners declined to take one. "You must take a basket," the East German stated, "Everyone must take one." That was the communist way. Everyone had to be the same, do the same things, and follow arbitrary but unnecessary practices. While shopping in a small but exclusive hobby shop later that day, I received a warm welcome after speaking to the proprietor in German and was invited behind the counter for a closer look at some miniature train models. A well-dressed East German entered the shop, obviously from the political or government upper crust. He glared at me—I was in my green dress uniform, as American soldiers were required to be in uniform while visiting East Berlin. The East German moved to step behind the counter himself, but the shopkeeper barred him and said that customers must remain in front. Not realizing I understood German, the East German asked why the American was back there. He then said that, I, the American "was a visitor and deserved a good reception." In a huff, the East German stormed out of the shop. The proprietor, to my great surprise, said in German, "A petty Party loyalist." It was an outspoken criticism of the system. I hoped the proprietor's actions wouldn't result in reprisals.

A final incident occurred while we were watching the Soviet guards at a memorial to Russian soldiers who died in World War II. Taking photos of official buildings and military facilities was strictly prohibited, but this memorial was a tourist site that allowed photography. Many in the crowd had their cameras out, strapped around their necks. Suddenly, a large motorcade with a police escort passed, and the crowd turned to see who was passing. It was Soviet President Brezhnev, General Secretary of the Communist Party, who had stopped in Berlin while on a trip out of the Soviet Union. While no photos were taken, a Soviet agent evidently thought that one had been. An agent invariably tailed us about East Berlin because of both my status as an American field grade officer and the War Plans assignment. This was undoubtedly known because of the Russian agents often seen observing the American headquarters in Heidelberg. The U.S. had similar liaison personnel observing senior Soviet planners, and what they were interested in doing and seeing.

Later that day, we ate at the Moscow House, a famous East Berlin restaurant. Parking the car, we carefully placed all valuables out of sight and locked the doors. After a fine meal, we returned to find a passenger door unlocked. All possessions were in place, and everything appeared to be in order. We assumed that we had somehow overlooked the one door. The next day, after taking many more photographs, Nannette remarked that the film might have slipped off the sprocket, as the exposure counter was above the number of shots expected. We opened the camera, only to discover that some clever agent had popped the lock on the car door, located the camera under the car seat, removed the film, and advanced the film counter to where it had been before returning the empty camera. It was just another example of the communist obsession with control and knowledge of everything.

▼ ▼ ▼

The friends from West Point continued to meet and socialize the first half of 1981. Larry Ellis moved from Heidelberg to the 8th Infantry Division a few miles to the west, where he served as an infantry battalion executive officer. Harry Lumpkin continued working as a logistician with the Theater Air Defense Command, until receiving orders to attend the Armed Forces Staff College in Norfolk, Virginia, that Major Ellis had attended two years prior. When the three of us departed West Point, Harry had come directly to Germany, so his three-year tour ended a year before ours. We and the Ellis family farewelled our friends, certain that we would encounter them again.

The end to the Jordan family's three years in Germany came more quickly than any wanted. I had grown professionally and come to truly understand the value of allies. American strategy depended on having allies around the world, and those officers who had neither the experience with nor the appreciation of allies were not prepared to advance further. The family had seen Europe and gained an appreciation of things beyond the narrow world of most Americans back home. Though living in the greatest nation on earth, not having an appreciation of the rest of the world's riches was a tremendous shortcoming. The entire family had seen firsthand the face of the communist adversary, and better understood why the Army was such a taskmaster. Even at their young age, the boys seemed to understand. Before, the Army was just Dad's job, with neat things like tanks and rifles. Afterwards, it was the thing that helped keep us and our allies free.

In 1982, we learned of our next assignment. Following my job as War Plans Officer in Europe, I was to work in the Pentagon, or as it was referred to by many officers, "The Building." We were headed back to northern Virginia. Unlike the first trip there to attend Amphibious Warfare School with the Marine Corps, this trip felt like going home.

CHAPTER 12

The Building

PRIOR TO RECEIVING orders to move, Armor Branch had hinted that a likely follow on assignment as a War Plans Officer in Europe was a job in the Pentagon. The family had traveled back to the DC area a couple times in the last year, once on leave and another time for a meeting to discuss supporting a peacekeeping mission in the Sinai Peninsula. During spare time we had toured various neighborhoods and collected dozens of floorplans for new homes being built in various developments. Over the next few months, we pored over the floorplans trying to decide on our dream home.

When official orders finally arrived, they fortunately coincided with another trip to attend a meeting at the Pentagon. My personal mission was to go to contract on building a new house in a growing community near Woodbridge, where we had lived while studying with the Marines. We had decided on the Montclair community in Dumfries, Virginia, which was a former gated community that no longer limited "undesirable" residents. We would later learn that it had basically been impossible for people of color to buy there prior to the Civil Rights Movement. Now, Montclair was greatly expanding with a more diversified population. Still, economic status tended to be uniform—primarily middle-class professionals lived there—due to the cost of the homes. As the boys were in school, Nannette could not accompany me to provide the normal wifely oversight. Armed with a precise list of house requirements that Nannette and I had agreed upon, I was to select a lot and ensure the contract covered all our needs and desires. I had joked about simply being the messenger carrying the requirements to the builder, but I secretly enjoyed the confidence she had placed in me. The construction boom was such that five months was enough time to complete the home, in time for our arrival from Germany. Once the contract was signed in January of 1982, we would not see the house until returning to the States permanently. Fortunately, our friends the Lumpkins had bought a house about four miles north of Montclair and provided reports and photos of the construction process. Harry had finished his time at the Armed Forces Staff College and been assigned to the Army Staff in the Pentagon. As always, good friends made a tremendous difference; they often acted as surrogate or extended family. That was especially true for Army families.

The final flight back to the States was nothing like the cargo plane ride three years earlier. We enjoyed adjacent seats in a comfortable wide-bodied airliner. Landing at Washington Dulles International Airport, we rejoiced at being home and closer to family. The experience in Germany had been enjoyable, but we were ready for the next adventure, which included our first time building and owning a home. Plus, chances were, we would see Germany again.

Parking in the driveway, we saw our new home for the first time. It was all that we had imagined. In actuality, the house was an average sized center-hall colonial, a style ubiquitous in Northern Virginia. But with three bedrooms, a living room, a dining room, an eat-in kitchen, a family room with a fireplace, and a full-but-unfinished basement, it seemed perfect for us. We soon had a nice, elevated deck built, that was accessed from the family room. There was a patio area underneath the deck that could be reached from the basement via a sliding glass door. Our home was located on a cul-de-sac, which offered outdoor space for the two growing boys and the host of kids who lived in houses on our street. It seemed that every third or fourth house had a military family. Several Marine officers, active or retired, lived there due to the proximity of Quantico Marine Base. Army, Navy, and Air Force were also well represented. On one side of our home was an Air Force couple with two daughters. On the other side was an Army couple with two daughters about Larry Reginald's age and a son close to Karl's age. As fate would have it, the father was an Army military police officer who worked in the office next to mine in the Pentagon. Joe Hebert was genial and outgoing, typical of most young majors. We became friends and carpool mates.

The Pentagon was a virtual small town unto itself. It was easy for someone not familiar with it to become hopelessly lost. Fortunately, I had visited a few times and quickly learned to navigate its 17.5 miles of corridors. The five-sided building consisted of five concentric rings of offices with narrow open areas, perhaps 20 feet wide, separating each ring from the adjacent rings. With five rings, each five floors high, The Building was the largest office building in the world. There were several radiating corridors that connected each ring to the others, so movement throughout the building was facilitated—provided you understood the layout and complex room numbering system. In addition to the extensive corridors, it was 6.5 million square feet, covered 29 acres with an additional 67 acres for parking, and housed a total workforce of over 25,000. There were also multiple basement levels that did not follow the floor pattern above ground. With its confusing maze of passages, the apocryphal myth was that an Action Officer had become lost down there when the structure was first occupied in 1943 and was not found until after the war ended in 1945.

The Building was practically self-contained. In addition to offices, workspace, briefing rooms, communications centers, and auditoriums, there were medical and dental clinics, chapels, dining rooms, snack bars, a gymnasium, and even an area with banking facilities, retail shops, barbers, a pharmacy, bakery, and shoe repair. The area's mass transit authority, Metro, had light rail and bus terminals located there, and a helicopter pad allowed priority visitors to alight steps from an entrance.

Upon reporting to my new duties in the Pentagon, I was assigned to the Office of the Deputy Chief of Staff for Operation. The head of Deputy Chief of Staff for Operations (DCSOPS) was a three-star general—though many one- and two-star generals ran the different parts of the organization. I was in Strategic Plans and Policy, while my neighbor Joe Hebert was in War Plans. Having a next-door neighbor who worked in the next office in the Pentagon was a most unlikely but fortunate occurrence. It made carpooling a natural fit.

Unlike the War Plans office in Europe, which developed and maintained actual combat plans, Joe Hebert's office was responsible for coordinating Army-level support for those joint operations. Strategic Plans and Policy, where I worked, coordinated the development of Army-level strategic guidance, usually in documents that allocated forces from all the services. I was also to be the Army Action Officer in developing something called the Unified Command Plan (UCP). The UCP established the major joint Combatant Commands—like U.S. European Command, U.S. Pacific Command, and the Strategic Air Command—and designated the geographic areas for which each was responsible.

I learned an interesting political lesson while updating the UCP. During this Pentagon tour, the UCP was expanded to create the U.S. Central Command, or CENTCOM. CENTCOM was to be responsible for the Middle East Region, where Operation Desert Storm would later be fought. Representatives from the Joint Staff and each service worked collaboratively to design the new command, and I experienced the intense inter-service rivalry that takes place concerning the allocation of resources. U.S. Central Command was seen as a ground theater that could favor the growth of Army force structure, and therefore the growth of the Army budget. Since the Defense Budget was often viewed as a zero-sum game, growth in one area could mean reductions in others—and the Navy in particular saw itself as a potential bill-payer. In designing the Area of Responsibility for the new command, the Army, Marines, and Air Force argued the waters around the landmasses be included. The Navy opposed it, as it would increase the maritime area of a land focused command. Plus, any naval forces operating in those waters would fall under a land commander's command rather than a Navy admiral's. Fortunately, the strategic and operational arguments to include the waters adjacent to the land areas in the AOR prevailed. That decision would prove critical to success in later conflicts.

In a similar manner, which command would be responsible for Alaska was being argued, this time by political as well as military players. A senator from Alaska had worked ceaselessly, as most senators do, to ensure his state benefited from defense presence and construction. He worried that Alaska would fall under the organization responsible for the lower forty-eight, becoming merely one of forty-nine, and certainly less populated by the military than some other states. The senator saw the best solution as a separate joint command for Alaska, something the services saw as neither necessary nor efficient. For the Army and Navy, there were reasons to link Alaska and Hawaii, which were important to U.S. presence in and around the Pacific Basin. But the Air Force sought to keep Alaska tightly linked to mainland North America for air defense purposes. The senators from Hawaii and Alaska helped resolve the impasse by concluding that having both states in the Pacific Area of Responsibility made sense. The Joint Staff briefing, supported by the Army and Navy, gained those senators' endorsement in the Senate Armed Forces Committee. Without that endorsement, the services might have had even more difficulty reaching agreement. The Air Force was assured that the North American Air Defense Command would effectively incorporate Alaskan airspace, while the Army and Navy gained the Pacific Basin linkage they sought. Through this experience, I learned how important it was to seek solutions that not only delivered the desired results but also avoided significantly disadvantaging any stakeholders.

The Jordan family enjoyed the many opportunities offered in the DC area. We took monthly weekend trips to visit the many museums, cultural centers, tourist sites, and other venues. Preferring more unstructured, carefree, and play-oriented activities, the boys called these excursions "Forced Family Fun" after Nanette and I had jokingly mentioned that we would be having some family fun weekends, even if they had to be forced.

The boys rejoined old friends along with making new ones. The Lumpkins and their two sons lived only a few minutes away; Major Rick Rhodes and his family had also moved to the area at the end of their Germany tour. Larry Reginald and Karl began to understand that everywhere you moved in the Army, you usually found friends that you had known somewhere else. This, plus their natural enjoyment of new adventures, prevented them from resenting picking up and moving. Civilian friends sometimes asked Nannette and me how the boys adjusted to the transient Army life, but they never appeared fully convinced when we told them the boys "loved it." Our sons also began to realize that not everyone had traveled as much as they had. This was driven home for twelve-year-old Larry Reginald when his class was given an assignment to present oral reports on various countries. He excitedly selected France because he had visited Paris with his parents. To illustrate his report, he showed some photos he had personally taken of the Eiffel Tower. To his frustration and anger, some of the children said he hadn't really seen the tower because "it's in France and nobody here goes there." Others teased him and said he was bragging. Most of them were from civilian families, and international travel was not something they had done or considered. We had to explain how lucky he was to have gone places many could only read about. We also explained that there would always be people who were envious of the things that others did, and not to let that stop him from appreciating those things.

The first few months in our new home also saw an addition to the family. While in Germany, Karl had repeatedly asked for a dog. That was not practical in Germany, especially with the limitations of stairwell living. We had promised to get him one once back in the States, thinking he might forget and move on to the next desire. That idea evaporated when Karl's teacher brought in a litter of puppies and offered them for the taking. With our approval, he was able to provide a forever home for a small, calm, brown male with a black-colored area on his face. Karl said the puppy looked like a bandit with a mask. The name stuck, and Bandit became part of the family.

Professionally, seeing how the highest headquarters in the Army operated and integrated with other agencies was again like graduate-level schooling, this time in strategy. Some officers proudly boasted of having never served in the Pentagon, preferring to be "field soldiers." I had been told, and now realized, that understanding how the Army made policies and decisions would assist a commander in implementing those policies and decisions in the field. It was also a necessary gate to higher responsibility and rank.

The prospect of higher rank became a reality early in 1983. The lieutenant colonel promotion list was released, including the name Larry Reginald Jordan. While excited, I realized that I still had over two years remaining on my three-year Pentagon assignment. The promotion and pay raise would be welcomed but it would not necessarily expedite my return to troop duty. As much as I appreciated all I'd learned in the War Plans position in Germany and was now learning in the Pentagon, my true desire was to lead soldiers. Again, the prospect of doing that arose as the annual selection board for battalion commanders convened in February.

My good friend Major Ellis and I both eagerly awaited the opportunity to command. Having just been selected for promotion to lieutenant colonel, with actual promotion probably a year away, I didn't really anticipate being selected for command that year. But it seemed our wishes had been granted when we found both our names on the battalion command list as promotable majors. Unlike later when email and other digital communications were the norm in the Pentagon, the list was posted on a bulletin board outside one of the personnel offices, aptly nicknamed "The Wailing Wall." No one is reported to have ever seen lists being pinned to The Wailing Wall. Promotion and command selection lists seemed to magically appear. But news of such postings traveled around The Building like wildfire, drawing crowds to either rejoice or wail in disappointment.

▼ ▼ ▼

To my surprise, I was allowed to depart from the Pentagon after completing barely a year of my thirty-six-month assignment, receiving credit for an Army Staff tour. I actually felt that I had probably learned all that was important, and the remaining time would have been repetitious. I said something to myself that my friend Larry Ellis and I would often say when discussing whether we would progress professionally: "Someone has to do it; might as well be us." For me, battalion commander was the job of my dreams.

CHAPTER 13

The Job of Dreams

THE CROSS-COUNTRY TRIP to assume the job of my dreams in Fort Hood, Texas, included the usual stops to see family. We spent time with Nannette's family in Lexington, Kentucky, before heading to battalion command, where getting away for extended leave was doubtful. Top Owens and I enjoyed swapping old Army tales, some about Germany, others about Fort Hood—and, naturally, some tall tales about Vietnam. Nannette's dad, the retired First Sergeant, had started working for the Federal Bureau of Prisons at a minimum-security facility for non-violent offenders. He joked that twenty years of dealing with young soldiers, especially "knuckleheads," prepared him to handle convicts. Actually, like most NCOs and officers, he loved being around and leading soldiers. He greatly missed the Army, as nearly all who serve a full career do upon retirement. He took the same care and interest in counseling and helping the often-young inmates as he had with the sometimes young and immature soldiers under his leadership.

Nannette's parents had enjoyed their time in the Army and kept in touch with friends they'd served with. One incident in Germany three years earlier had shown me how lasting some friendships can be. Nannette's parents had visited us in Germany and borrowed a car to drive a few hours south to Baumholder, where they had lived twenty years earlier. While walking along old familiar streets a German woman stopped, stared at Nannette's Mom with wide eyes, and cried "Lillie!" The woman had known the family and remembered the kind, amiable American wife. Friendship can be timeless and knows no national boundaries.

This trip to Texas had no stop in Kansas City. Dorothy Jordan had long retired from her job as a social worker and finally gone to live with daughter Joyce in Illinois. Medical issues had slowed her physically, but not mentally or in spirit. I increasingly admired the World War II generation. They were survivors. They had endured and survived the Great Depression and the hardships of World War II. Those who stayed home and worked to keep the country running sacrificed as did those who went abroad to fight. I thought about the old ration stamps for butter and fuel I had found in old furniture drawers as a kid; the work my mom had done in a Defense Plant in the evening after teaching school in the day; the way she spoke proudly of "surviving hard times." I did not see that same spirit among young families today—with one exception. Military families seemed to have that same grit and ability to face down any challenge. I realized that my own family had grown into one that I had seen and admired when I first met Nannette's parents.

▼▼▼

Arriving at Fort Hood really was like going home. Nannette had been there with her parents, I had been there for a brief time as a cadet and later as a lieutenant, and we had met and married there fourteen years earlier. Much about the huge post had not changed, but a few significant things were different. The 2nd Armored Division, Hell on Wheels, was exactly where it was previously. The 1st Armored Division, Old Ironsides, had moved to Germany, and the 1st Cavalry Division had returned from Vietnam to take its place. While designated a cavalry division, it was organized like all other armored divisions. The new addition to the post was a separate air cavalry brigade. It was organized around attack helicopter squadrons and assault helicopter and cargo squadrons and battalions. The unit worked for the Corps, and its headquarters was located on the street directly across from the headquarters of the 1st Battalion, 67th Armored Regiment, which soon-to-be Lieutenant Colonel Jordan was to command.

A place as large as Fort Hood had a tremendous amount of family housing. We were pleasantly surprised to learn some relatively new housing, Comanche Housing Area, had been built on the western edge of the cantonment area, or main portion of the post. West and north of Comanche were maneuver training areas and gunnery ranges. Between the new housing and that located on main post proper, there were sufficient quarters for perhaps 60 percent of married officers and priority housing for battalion commanders. The only problem was that I would not be a battalion commander for two months; I was still a promotable major, which meant we needed to go into temporary housing.

Arriving families often spent a few days in the on-post transient lodging, which usually resembled small hotels or motels. In Fort Hood at the time, it was a wooden WWII-era single-story block of units. Each unit had two small bedrooms, a kitchenette, and a small area that served as living and dining room. Shocked, Karl proclaimed it a "slum place." The family had a good laugh at his sincere outrage, then and for years to come. Three nights in the "slum place" was all that any of us wanted, so we explored the surrounding communities for a rental. That made sense, as not only was the authorized housing that accompanied my becoming a battalion commander two months away, but waiting for a set of quarters to become available might mean a longer wait.

We found a suitable house in Killeen, Texas, in an area that had yet to exist when we left in 1969. The nice single-story ranch style house on a quiet street seemed perfect. It even had a fenced in backyard for Bandit. We briefly considered whether to remain in the house if offered on-post housing but quickly dismissed the idea after thinking about the drive to work and the distance to post facilities. Plus, while a nice community, it was not a typical military community, despite the Army families sprinkled throughout.

One unusual event highlighted the difference between that neighborhood and the environment we desired. Bandit was almost two years old, house broken and obedient, and several neighbors had commented on him being a handsome and well-behaved hound. He had never sought to run away, appreciating the play, attention, and regular meals received at home. But one afternoon, Bandit disappeared from our backyard. At the time, I was traveling to attend some required schooling. Fearing Bandit had somehow been attracted by another animal, Nannette searched the neighborhood, calling his name, which he always responded to. Two days passed with no results—other than the boys believing they'd heard him barking in one of the houses close by. The family repeated their search, loudly calling his name. Again, they heard barking. In elevated voices, they discussed calling the police to help locate the dog. Amazingly, Bandit showed up at our door that evening. Someone, who would rather avoid police involvement, released him.

After completing the Pre-Command courses at Forts Knox and Leavenworth, it was time to take command. Fortunately, quarters on post became available before the big event. With our new home established, we were prepared for the next two years as the First Family of the 1st battalion, 67th Armored Regiment. The day of the change of command, we were invited to meet with the Commanding General, or CG, of the division prior to the ceremony. It was normal for the CG to meet and welcome the families of new battalion and brigade commanders, but this meeting had another purpose as well. Being promotable but still some months away from the actual promotion date, I was authorized by the

Army to wear the rank and assume the authority and responsibility of a lieutenant colonel, but I would continue to be paid as a major until officially promoted. This process was called "frocking." For all appearances, Major-promotable Jordan was now Lieutenant Colonel Jordan. Following Army tradition, Nannette helped pin the new rank insignia on my shoulders, as she would for every future promotion—except one. The boys watched, excited and proud. Years later, when the boys would be in their twenties, they confided in me about how deeply proud they had been then and throughout my career. They expressed that watching how their mother and I interacted with people, professionally and socially, had made an impression and helped shape their manner of living. It reminded me of the lessons learned from family as I was growing up.

The pace of training and other activities was as fast—if not more so—than that as a platoon leader at Fort Hood, a company commander at Fort Riley, or a battalion executive officer in Kirch-Göns. At the time, the battalion was one of perhaps half a dozen Army battalions equipped with the new M1 Abrams Tank. More would eventually be fielded. The new tank's improved daylight and thermal night sights, stabilized gun for firing while traveling, and other advanced systems substantially improved gunnery accuracy. Tremendous amounts of field maneuver training took place to master the greatly increased tactical benefits of the Abrams Tank's increased speed, power, and maneuverability.

The opportunity to demonstrate the battalion's capabilities arose quickly. Each battalion had annual evaluations that used the formal Army Training Test of various tactical missions—defend, attack, movement to contact, and so forth. A team from the division headquarters graded the tested unit as it operated against an OPFOR, or enemy Opposing Force, which was usually another battalion in the brigade or division. Often for the 1st Battalion, 67th Armored Regiment, the OPFOR was our sister battalion in the brigade, the 3rd Battalion of the 67th. Though we shared a friendly rivalry, I got along with the other commanders, all of whom I would fight alongside if we ever went into combat. Inside the battalion, we referred to our sister battalion as "Brand X."

I never skipped an opportunity to train the battalion. If another unit for some reason could not or did not want to use its reserved training area, I quickly volunteered. If I happened to learn of a vacant area, I would sometimes issue a Fragmentary Order, or FRAGO, for one of the tank companies to conduct a no-notice move to "occupy the area and conduct operations." This was to see how quickly the unit could prepare and move from the motor pool to the field area using proper tactical procedures. Obstacles the Engineer Battalion built as part of their own training offered a similar opportunity. I would request that they leave the obstacle in place, rather than restore the terrain to its original condition as was called for. I would then issue a FRAGO for one or more of the companies to attack and breech the obstacle while another defended it as a small OPFOR. Afterwards, I gladly restored the terrain in return for a great extra training opportunity.

Extra training was always at a premium in Fort Hood. Even with the vast military reservation, a tight schedule was maintained to allocate maneuver space for the two divisions and many other units. Compounding the demand for maneuver space was the Army's decision to maximize both weekday training and weekend free time for soldiers. That was welcomed by all, including families. Weekend training was only authorized when the entire brigade or division had extended training over a weekend period, or by special permission of the CG. I received such permission by proposing to compensate soldiers with extra time off the following week. After some initial grumbling, the soldiers embraced the idea. They could maneuver freely and widely across the training areas, unincumbered by sharing the land with others. Additionally, if they went to the field Thursday through Sunday, then worked cleaning and repairing the vehicles Monday through Wednesday, for example, they had the next four days off. This appealed to soldiers and families, as they could travel or pursue other activities without taking precious leave days. It also appealed to the psychology of soldiers. While everyone else in the division was working on Thursday and Friday, they had off. It didn't matter that everyone had the same number of days off across the month. What mattered was that they were off for four days in a row while their buddies elsewhere were working. That was an interesting lesson on what motivated some people.

I also found interesting ways to train the special platoons, including the Battalion Scout Platoon. Remembering my Scout Platoon days in Vietnam, I selected proficient and self-confident officers and sergeants to lead it. They had to have the initiative to operate independently and with minimal supervision. Sometimes when Brand X was training in the field, I would give the Scouts the mission to surveil their movements. They were to report how well the observed unit performed various tasks, what routines or habits they exhibited, and which companies did which tasks well or poorly. I wanted them to learn to observe without being detected and to report important details. Along with training the Scouts, I was gathering important intelligence about my potential OPFOR for the next evaluation's mock battles. That paid off several times.

Aside from the annual Army Training Tests and semi-annual tank gunnery, there were many other subjects and tasks to master, including qualifying soldiers on their individual rifles or pistols; training medics on first aid and combat lifesaving; and training cooks, clerks, mechanics, and others on their particular skills. However, the first year's largest event was the chance to participate in a REFORGER Exercise. It had been two years since leaving Germany, and I looked forward to the maneuver and seeing the country again. Should there be free time, I would not object to a good meal of Schnitzel and a frosty mug of German beer. A nice German wine, either a Kabinett or Spätlese, would also go down well.

The training was intense, and I focused subordinate leaders not only on tactics but also on the process of drawing and turning in the equipment, which had been critical during past REFORGER experiences. Maneuvering across the German countryside brought back many memories; more importantly, it allowed the battalion to practice what it needed to do if called upon to fight. At home, when addressing the officers, the NCOs, or the entire battalion, I would often say, "Work hard at what you do. There could be a Russian regiment in your future." A Soviet regiment usually had three battalions of either tanks or mechanized infantry fighting vehicles, similar to but less capable than their U.S. counterparts. The problem was that the Soviet Union had many more of each. The battalion was to fight, stop, and destroy a Russian regiment three times its size. Plus, I needed to retain enough combat power to continue to fight, because a second Russian echelon of equal size would not be far behind. Should that day ever come, I was glad that we had superior vehicles, good comrades on our flanks, artillery and attack helicopter support, and the best air force in the world to provide close air support. The practice combat on REFORGER went extremely well, as did the turn-in of equipment. Remembering my cocky remark to the battalion commander during my first REFORGER about sending any tough tasks to my company, I chuckled. All my company commanders were certainly doing better than I had done back then, and I told them so. The time spent in Germany was enjoyable, as was returning home to family.

One memorable incident occurred early in the exercise while the battalion was drawing the equipment from warehouses. An American film crew preparing a segment for *60 Minutes* arrived and asked to film the activity and get comments from the commanding officer. The division public affairs officer had briefed all commanders back at Fort Hood on how to handle interviews and respond to the media. The guiding principles were to be friendly but professional, factual, and non-controversial in comments. After a few seconds of filming the still frenetic activity, the interviewer asked me what was occurring. I explained the tasks and that the soldiers were doing a great job. The interviewer then asked, "With this display of the U.S. ability to get troops here quickly, does this encourage the Soviets to use nuclear weapons?" That controversial issue was the last thing that I had expected, let alone was prepared or eager to discuss. I replied, "I'm not an expert on nuclear policy, and don't know anything about Soviet plans. However, as you can see, these soldiers are doing a great job that should make Americans proud." With that, the film crew's smiles disappeared. The interviewer gave the "cut" signal to the cameraman, and the crew departed without another word.

After reporting the interview to the public affairs officer and the brigade commander as required, I reflected for a moment. I had just practiced something that would remain with me throughout my career: Treat the media with respect; tell them the truth; stay in your lane and do not speculate about things that are outside your awareness or responsibility. Later, I would learn how to hold media representatives accountable for the accuracy of their reporting. While many in uniform distrusted or disliked the media due to the negative coverage in Vietnam, I tried to be philosophical. The founding fathers had considered a free press essential, with which I agreed. However, I also thought there were good, professional media representatives, and bad, unsavory, callous ones. The key was to recognize the difference, and how to respond to each.

After returning from REFORGER that autumn, the battalion had a short time off before training for the next major event. This would be a rotation at the National Training Center, or NTC, at Fort Irwin, California, in the Mohave Desert. NTC was the toughest evaluation for a unit and its commander at the battalion level. The three-week evaluation included maneuvers against the permanently assigned OPFOR there: a U.S. Army brigade trained in Soviet doctrine and tactics, using surrogate Russian equipment, and purported to be the "best trained Russian regiment on earth." They routinely ran roughshod over evaluated units, called "Blue Force" after the blue color used to depict friendly units on an operations map. Enemy forces were depicted with red, and thus called "Red Forces."

The training had the predictable and enjoyable break for the holiday season. Army tradition was to have great feasts in the dining halls for Thanksgiving and Christmas. As mess sergeants tried to outdo each other, scrumptious meals and decorations were the norm. Families could also join in the festivities for a very modest fee that covered the government cost of the food. It was often an upscale restaurant experience at a less than fast-food cost. Unit commanders routinely showed up to praise and encourage the cooks toiling late into the night and early morning to prepare the next day's spread. At Thanksgiving, some volunteered to help on the serving line that soldiers and families filed past. Again, this was an old Army tradition.

The other break in the routine during spring was a ten-kilometer road race held annually in Austin, Texas, called the Capitol 10-K. With running being a daily activity for many at Fort Hood, both individuals and some complete units entered the race. At the time, the Army had yet to issue a standard PT uniform, except for some reversible black and yellow shirts to designate teams for competitive physical activities. But the 1st Battalion, 67th Armored Regiment had purchased distinctive PT shirts for each soldier from the unit's petty cash fund, which was supported by a small allotment from the Army for morale purposes, donations from the battalion's personnel, and, more importantly, bake sales and other events conducted by unit spouses. The shirts were bright Armor yellow with a large battalion crest on the chest and the name of the battalion emblazoned above it. With those shirts, it was impossible to mistake which unit was passing during morning physical training, which the soldiers loved, as they sang Jodie calls or made offhand remarks about the lesser dressed troops. In turn, others jeered the brightly clad troops—while silently wishing they had distinctive shirts, too. It would be a few years before the Army approved and fielded a distinctive PT uniform that identified the wearer as U.S. Army, in which soldiers could take pride.

Each year, the battalion took nearly all its soldiers to run as a unit in the Capitol 10-K. The public was ecstatic to see 500 soldiers running in step in a mass formation, all wearing bright shirts with their battalion colors and the national flag in front. The battalion had a reputation at home as well. The division commander was an avid runner and held monthly five-kilometer races on post where company-size units and individuals could enter and compete. The battalion's Company D, or Delta Company, won the company competition twelve straight times before retiring the trophy. I won the over-thirty individual bracket a couple times. While I enjoyed running, I did not fool myself into thinking I was a quality distance runner.

The training accelerated as the NTC rotation approached. Only one other M1 Abrams tank battalion had ever been to the NTC for evaluation. That battalion was also in the 2nd Armored Division, and, as some joked, had only used Abrams Tank's speed to get into the kill zone quicker, where they were declared destroyed. They had suffered the same trouncing that Blue Force units routinely received, despite the high expectations many had for the newly modernized battalion. However, that was how units and commanders learned. The After Action Review, or AAR, following each mission covered every action and decision in great detail. Commanders were required to describe what they intended to do and why, then discuss what happened. However, Blue Force commanders sometimes didn't have a clear picture thanks to the confusion of simulated battle, so both the OPFOR commander and the umpires, or Observer Controllers, described what actually occurred. As in actual combat, the fog of war often made complete understanding of any situation or action impossible. How well a commander worked through that fog often determined the outcome.

Energized by the knowledge that we would be the second ever modernized tank battalion to go to NTC, we were determined not to suffer the same fate as the first. That would reflect on not only us but the tank and the entire Armor Force. The months of hard training paid off, but not without a few setbacks and the normal shortfalls which enhanced the training value. My partner unit was the 2nd Battalion, 41st Infantry, commanded by Lieutenant Colonel Jim McDonough. Jim was a year behind me at West Point and had been a brigade boxing champion in the lightweight class at the Academy. He was a tough and competent battalion commander and the two of us worked well together. During the force-on-force portion of the rotation we exchanged ideas and coordinated actions when both fighting the OPFOR. For example, we both did things that intentionally caused the OPFOR to take notice late at night, so that the OPFOR commander might be awoken for guidance in the middle of the night. The theory is that a tired commander is a dumb commander—he or she will make mistakes due to being too tired to make good cognitive decisions. Some Blue Force commanders tried to power through the exercise on minimum sleep, catching two or three hours whenever they could. By the third or fourth day, they were like zombies, as were their subordinates. Both McDonough and I had plans that ensured we got at least five to six uninterrupted hours of sleep unless something very unusual occurred.

The 1st Battalion, 67th Armored Regiment had the normal small successes and failures, but two major events set us apart. On the final force-on-force mission, the battalion had to defend a long and wide valley against an entire OPFOR regiment. Due to the Abrams equipped task force's capabilities, the OPFOR assembled the largest regiment to date to mount the attack. I had several challenges to overcome. The highly effective OPFOR scouts would undoubtedly observe my defensive preparations and pinpoint my locations. One of their tactics was to have dismounted troops infiltrate before dawn on the morning of the attack and destroy as many vehicles as possible with shoulder fired rocket launchers at first light. Then they would attack with massed tanks and other vehicles to punch through the weakened defenses.

To counter this tactic, I placed one company on the valley's western flank, ahead of the rest of my force, but where they could destroy the enemy passing down the valley toward the main objective. I knew the enemy had seen the company and that dismounts would infiltrate. I instructed the company to dig in, and to get everyone under cover inside the vehicles, take down radio antennas to avoid damage, and button up all vehicles by closing the hatches twenty minutes before dawn. At ten minutes prior to dawn, when the enemy infiltration was complete and they were awaiting first light to attack, I ordered an artillery strike directly atop the company using airbursts; this strike would rain shrapnel down on the entire area, but the worst damage to the tanks and infantry fighting vehicles would be chipped paint and beat up fenders. However, that shrapnel would decimate the enemy dismounted troops. The umpires assessed over a hundred enemy casualties, which completely disrupted their attack. The OPFOR commander complained bitterly but was told, "War is war."

The main defensive area was the only thing left. While I could not hide that I had built it and where, I used mobile trailer-mounted smoke generators to continuously obscure the area. I emplaced the strongest and most complex obstacle I could build, along with fighting positions for two or more companies to cover it—though I would actually only place one company there. However, what the smoke enabled me to construct without being detected by the enemy was a second line of fighting positions a thousand meters behind the first, with mine fields but no visible barbed wire or tank ditches. This line could also accommodate two companies, and I would put most of two companies there after dark. Finally, I placed one of those companies' platoons back on the enemy's final objective for a last-ditch defense effort, if required. My mission was to prevent the enemy from passing that point at all costs.

When the OPFOR regiment attacked, it was without most of their reconnaissance capability, much of which the pre-dawn artillery strike had destroyed. All vehicles, including OPFOR, were fitted with laser training devices for shooting and registering hits on other vehicles. You could shoot the enemy and register kills—or shoot friendly vehicles if you were not careful, with the same result. There was no arguing about who shot who. The system clearly showed the results. The company I had placed on the flank ahead of the main line killed half of the lead attacking battalion before being bypassed. When that attritted battalion hit the obstacle and the dug in Abrams company, it was quickly destroyed. True to Soviet doctrine, the OPFOR commander pushed his second battalion forward with the third following closely. The second battalion lost much of its force in the obstacle but managed to open breeches. Its remnants and the third battalion broke through the company I had placed there, only to run into the fire from the other two companies 1,000 meters away. When that fight was over, the OPFOR had bypassed that position with a single company and reported to his higher headquarters that the way to the final objective was open. Placing even a single vehicle on that objective would amount to victory for him. Jubilant, he started the four-kilometer dash to the objective. At that point, I unleashed one of the companies into the unsuspecting OPFOR's rear, and between that company and the platoon to their front, every remaining vehicle of the OPFOR regiment was destroyed. In my mind it was a pyrrhic victory—I had prevailed but lost over half of the command. I later learned it was viewed as an absolute success. It was never intended for a Blue Force battalion to succeed against such a force in that situation. The victory was a testament to the training and the soldiers. I gave them the credit, and only secretly enjoyed pride in my tactics.

The last portion of the rotation was a live fire exercise. A day and night defensive live fire exercise was conducted using electrically controlled targets that represented an advancing regiment. The undulating terrain would hide or reveal the targets as they were raised successively closer to the defensive position, similar to how actual vehicles would come in and out of view on a real battlefield. Just as in an actual Russian regiment, there were so many targets that many Blue Force units were overrun as the targets crested the last ridgeline. The secret was to destroy as many as possible while they were still far away so they did not reappear closer later in the exercise. With its superior optical and thermal sights and fire control system, the Abrams easily destroyed targets, and the daylight run was successful. That night, the weather took a turn for the worse. Winds were so intense that the targets would not function properly. The exercise that was supposed to start at 2100 hours, or 9:00 p.m., was finally cancelled at 0130, or 1:30 a.m.

Four and a half hours later, the battalion had its last mission: a live fire movement to contact. Much like the maneuver portion of the rotation, the battalion would move to seize an objective and engage enemy targets—this time with live rounds. Therefore, these were electrically controlled targets, not actual OPFOR vehicles. These targets would also have gunfire simulators, and the observer controllers could "kill" exposed or effectively engaged Blue Force vehicles using the laser training system. The exhausted battalion had an inauspicious start to the mission. The artillery battalion supporting us was to fire live ammunition as well, and that support would be greatly needed to accomplish the tasks. But when the artillery fired its check rounds, they fell dangerously close to our friendly position. Somehow, the artillery fire direction center and the guns were not on the same page. After considerable delay, I was told that the battalion must attack without any artillery support due to the problem. Under the best of conditions, to do so without artillery was imprudent, and would probably not occur in combat. However, this was an evaluation, and the NTC commander would grade how the unit and its commander reacted to the challenge.

As expected, it was a tough day. The enemy employed chemical weapons which required the battalion to wear the hot and bulky chemical defense suits, and the going was slow without artillery to help eliminate enemy strong points. Late in the day, it was obvious the battalion would accomplish its mission, given enough time. A radio call summoned me to the head controller's jeep. The one-star general was going to issue me a change-of-mission order. Everyone in the battalion knew what that meant—it was the end of the mission, and time to return to the cantonment area to clean the vehicles and prepare to go home. After a brutal last day to the grueling three-week rotation, everyone was ready for the order to end the exercise.

The brigadier remarked that the battalion had a tough day and there was no sense to continuing that fight, as it appeared we would eventually gain the objective. However, he then said, "Change of mission. Reoccupy the night defensive position from last evening. You have a night defense to fight." Returning to my tank, I wondered how the exhausted battalion would react—they just had an extra day added to the exercise. I put out a radio call to the anxiously awaiting company commanders, who expected to hear "End of mission. Return to base." Rather, they received word to return to the live fire defensive position. After what seemed like an eternity, but was only a few seconds, each replied, "Roger, understand. WILCO," meaning they received the order and were responding. The battalion went through the night live fire, this time without weather problems. The Abrams tank's superb night fighting capabilities enabled us to amass the highest score of any battalion to date. I wondered if it was the training or the soldiers' anger at being required to stay out another night. Either way, I was proud of them. We finished the NTC rotation on a very high note. The battalion had done well; the Abrams tank had done well.

The homecoming was great. I enjoyed being back with family members and the buoyed reputation of the battalion. Hands down, the battalion was considered the best in the division. Plus, the Armor

Branch, probably at the division commander's request, had approved my remaining in command beyond the normal twenty-four-month tour. As I thought about it, I also wondered how much the corps commander had played in that decision. Lieutenant General Ulmer, who had been my division commander in Germany, had been promoted and was now the III Corps commander. We had occasionally run into each other and exchanged short remembrances of the 3rd Armored Division. Whatever the reason, I was enjoying the extra six months in command. It was both a reward for my success and a significant indicator to selection boards when considering me for a promotion, further schooling, or another command. I made a mental note of another lesson learned: A reputation, either good or bad, can follow you.

Our family enjoyed the nearly three years at Fort Hood. Contact with our close friends was mostly by phone and letters, although we did visit the Ellis family at Fort Polk, Louisiana, where Lieutenant Colonel Ellis was commanding an infantry battalion. He enjoyed it immensely, but his family found the adjacent Army town of Leesville a social, entertainment, and shopping let down. With dramatically fewer soldiers and families than Fort Hood or Fort Bragg, where the Ellis family had lived previously, their new locale did not have the same number or variety of services and facilities.

Meanwhile, the Jordan boys continued to enjoy life. Larry Reginald, now a teenager, preferred to go by "LJ." He completed his freshman and sophomore high school years in Texas, where football was almost a religion. He'd started the sport in eighth grade, our first year at Fort Hood. We realized how serious the sport was taken there when going to his first away game, where we mistook the middle school's field and stadium for a junior college's. During LJ's sophomore year, his high school team played in the state playoffs at the Dallas Cowboy's stadium. He and his teammates took turns having their pictures taken standing on the Texas star painted on the field's AstroTurf. He couldn't have been more excited.

By eleven years old, Karl was beginning to look like he might catch up with his older brother in size someday. He had numerous friends in the neighborhood, including the sons of Mike and Cindy Suozzo. Major Suozzo was the battalion executive officer for a few days before going to the brigade staff. Our families became close friends, and later we became godparents to their newborn daughter, Susie. Twenty-four years later, we had the pleasure of attending Susie's wedding. Army friendships tend to be long term.

Just before Thanksgiving, and our third year at Fort Hood, sadness struck the family. My mother suffered a heart attack. I travelled on emergency leave to be with her. What initially appeared to be recovery and medical success turned around a day after I arrived. The strong woman who survived the tragic loss of her father, the Great Depression, World War II, and a broken marriage while seeing two children through college lost that battle we all eventually lose. The family gathered to lay Dorothy Elizabeth Johnson Jordan to rest.

Five months later, the extended command tour ended. In the future, when asked how I stayed in command for two and a half years, I would often say that being a slow learner, it just took more time. I felt that was the best answer, since anyone asking that question probably knew that high performance was the only way it could have occurred. To answer otherwise would have marked me a conceited egotist in my mind.

Like all changes of command, I found mine bittersweet. I had put so much into the battalion, as had Nannette. It was almost like an extended family branch. Later in my career, I would coin a saying about the good times spent and shared with others. Every great experience was "a special place, a special time, and a special group of people." You could return to the place, and even gather the people, but you could not return to that time—except in your heart and mind. In those two places you could have them always. For me, it truly had been a job of my dreams.

The change of command was sweetened by the fact that I had been selected to attend Senior Service College, often called the War College. The capstone professional education course, each service had a Senior Service College, and most Army officers attended the Army's at Carlisle Barracks, Pennsylvania. Only about 20 percent of officers starting as lieutenants are promoted to lieutenant colonel. Probably no more than twenty percent of lieutenant colonels command battalions—fewer attend War College. It truly was a high hurdle to clear. In what was beginning to be a personal trend, I did not attend the Army school, but rather the National War College, run by DoD, and located at Fort McNair, in Washington, DC, proper.

CHAPTER 14

Back to the Nation's Capital

Attending the National War College meant the family could move back into the home we had built in the Montclair community. Many of the neighbors were still there, while a few—like my former carpool mate, Joe Hebert—had moved. It was nice to be back in the DC area; as luck would have it, our close friends, the Lumpkin and Ellis families, were also there at the time. It did not take long for the group to begin regular gatherings again.

All services attended the National War College, along with several foreign officers and State Department foreign service officers, and a few members from other federal agencies. My assigned section of fifteen people had a diverse membership, including a senior Secret Service agent previously in charge of presidential protection. Other members included a couple foreign service officers, one of which was a future ambassador to Bosnia. That region would later play a role in my life.

The War College year mostly consisted of readings, lectures, group discussions and exercises, and writing about classic and contemporary issues. It was also a great time for learning through experiences and associations with folks in the interagency arena. One of the year's highlights was the steady stream of noteworthy diplomatic and defense officials and experts, both domestic and foreign. Even a few congressional figures spent time at the college, as did ambassadors from some countries of interest in U.S. National Defense Strategy. Being located in Washington, DC, the National War College tended to draw a more varied group of speakers than the service schools located elsewhere, all of whom were stellar. The speakers' willingness to engage in extremely candid discussions was in part due to the non-attribution, off-the-record policy at the college.

Part of the curriculum was a two-week trip, during which students visited different regions around the world in groups of around twelve. I traveled first to Greece and then to Turkey, finding both fascinating countries with rich histories and current geopolitical significance. It was also interesting that the two NATO allies were often quarrelling over age old animosities and things like sovereignty over certain islands. To maintain harmony, the students were asked to use their official passports for one country and personal passports for the other to avoid showing customs officials the other country's entry stamps. It was all a silly facade, as both countries knew the trip's itinerary.

That spring also brought the release of the yearly promotion list for colonels. Several of my War College classmates and I were on it. Again, it was an early Below-the-Zone selection. I skipped ahead two years this time but still had at least a year before my promotion date was expected. Upon graduating from National War College in June, I was assigned to the Joint Staff in the Pentagon. I looked forward to the assignment, as it would allow me to practice what I'd learned at the War College and from experience in Joint actions during my previous tour in The Building. The assignment also allowed more time back in our Montclair house, which we had reoccupied upon returning from Fort Hood. Though we'd owned the house for five years, we had only lived in it for around two years.

The Joint Staff job was again a classroom for continued learning. I was assigned to the J-7 Directorate in an office responsible for planning and executing large scale war games and exercises at the strategic level. These dealt with serious matters of national security, sensitive political or diplomatic issues, or real or potential major conflicts. They ranged from things like reacting to a U.S. space shuttle conducting an emergency landing in an unfriendly or contested area to general global war, including the use of nuclear weapons. Most, however, were smaller yet important scenarios much more likely to occur. I learned a great deal about interagency operations and how various agencies conducted business. It was truly preparation to serve in the Army's highest levels.

The most personally impactful thing that happened during my time on the Joint Staff was meeting the officer in charge of the J-7 Directorate, Major General Fred Franks. Though he hadn't taught me, General Franks had been an English instructor at West Point when I was a cadet. He was also an armor and cavalry officer who had served with the 11th Armored Cavalry Regiment in Vietnam and been severely wounded. Drive, determination, and an enduring love of the Army had enabled him to remain on active duty despite losing a leg below the knee and requiring a prosthetic. A truly brilliant officer, he had gone on to command the 11th Armored Cavalry Regiment in Germany along the West German border. Clearly, the Army's most senior leadership did not see any professional or warfighting limitations due to his injury.

Early during the J-7 assignment, I had the opportunity to work on a special project for General Franks. Franks asked me and another officer to conduct some research and develop a concept for a briefing the general was to give. The other officer, touted about the Directorate as a fast-rising star headed for greater things, considered the assignment a trivial matter, undeserving of much effort. Franks had described the task as not urgent and instructed us to devote whatever time and effort we could without interfering with our regular duties. The other officer took that position, while I produced a concept and short briefing that satisfied the general. Franks thanked us both, and we thought that the end of it. Later,

I came to believe that General Franks had used the task to see if one or both of us would take the easy way out for two reasons. First, I made that assumption after seeing that he easily had the knowledge and skill to have accomplished the task in half the time it had taken me and the other officer. The second had to do with a job offer General Franks would make me at a later date. Again, lessons were learned: Consider every mission important; do more than your boss seems to expect.

Two additional significant events occurred in short order. First, I was finally promoted to full colonel. Lieutenant Colonel Kelly, my battalion commander at Fort Riley, was now a lieutenant general and serving as the Joint Staff operations officer, the J-3. I invited him to officiate at the promotion ceremony, which he graciously accepted. Time and increased rank had not diluted Tom Kelly's sense of humor. During the ceremony, as Nannette was preparing to pin one of the silver eagles of a Colonel on my shoulder epaulette, Kelly told her to ensure the eagle's head was facing forward so "Jordan can tell his head from his backside." A hardy chuckle passed through the crowd assembled to watch.

The other occurrence was the release of the annual brigade command list. Again, both Larry Ellis and I were on that list. In a congratulatory call between us, the oft repeated comment was made, "Someone had to be selected; might as well be us!" My excitement over the selection was short lived, however. Unlike the Army Staff, which released me early to take command of a battalion, the Joint Staff wanted me to complete a full tour before being released. I would be deferred until then. Armor Branch had tentatively slated me to go to the NTC as commander of the OPFOR brigade, probably because of my successful rotation there as an Abrams Tank battalion commander. I had been excited about the tentative assignment, while Nannette was prepared to make the best of it as a faithful Army wife. She was not excited, however, about Fort Irwin's very remote location in the Mohave Desert, the lack of access to local shopping or other facilities given the nearest town being several dozen miles away, and the notoriously extreme heat three-quarters of the year. There was also the extreme distance from family, but we had faced that before. The question of whether Fort Irwin would have been "home" to the Jordans was moot. With the Joint Staff decision not to release me, I settled in for a long wait.

Less than a month later, Major General Franks was announced to command the 1st Armored Division in Germany. I traveled to Brussels, Belgium, for a meeting at NATO headquarters a few days later. One night in my hotel room, I was awakened by a call from General Franks. Though the middle of the night in Europe, it was still working hours in Washington, DC. I immediately wondered if I had somehow made a major error, like leaving a classified document unsecured in the office. Instead, General Franks asked if I wanted to throw my hat in the ring for a chance to accompany him to Germany as his Division Chief of Staff. This would be a dream assignment, typically reserved for colonels who had already successfully completed brigade command and were considered prime for promotion to brigadier general. I had been a colonel for barely a month and had yet to command. I quickly replied that I would love to accept the job, but the Joint Staff had recently declined to release me. Franks opined that he thought it could be worked out. I remembered the "test" he had given me and the other officer. Undoubtedly, he'd been alerted to the possibility of heading to division command and was considering people to take with him.

To secure the release to head to Germany, I was directed to the three-star Army general who was the director of the Joint Staff. A gruff and no-nonsense officer, he asked why I wanted to be released after only fourteen months. When I explained, he said that he could understand the desire but did not have the authority to grant the exception. It would take the Secretary of Defense to do that. I left his office in stunned silence. How would the SECDEF, as he was called in shorthand, respond to such a relatively minor issue, considering the global responsibilities and challenges he dealt with daily? Would he even see a brand-new colonel for such a request? With more than a little trepidation, I walked into the SECDEF's outer office to attempt to schedule five minutes to submit the request but was intercepted by an Army brigadier general who served as Executive Assistant to the Deputy SECDEF. Upon hearing my issue, he said that it would not be necessary to bother Secretary Carlucci, and that the Deputy, Secretary Taft, could authorize the release if I made a clear and persuasive argument. The interview was scheduled and produced the desired result. I was not only released but granted credit for a full Joint tour. Given the mischievous grin on the brigadier's face when I departed the office, I wondered whether General Franks and the brigadier had worked it out beforehand.

Preparing to move followed the familiar pattern, with a notable exception—the move would coincide with LJ's entering college. The college application process had taken much of the spring. LJ had become a standout defensive back and running back on his high school team. Like his team in Texas, this one also went to the state finals. Several colleges and universities sought to recruit him, including West Point. We had never pushed him to select the Academy, yet he applied for a nomination. Both the Academy football staff and Admissions Office thought LJ was a good, qualified student and a fine athlete. But the football staff wanted him to spend a year at the Preparatory School to grow "bigger and stronger." LJ didn't wish to do that, and selected Wake Forest instead, which had offered a full scholarship. Wake Forest turned out to be a great institution, and he received a fine education. It appeared to us that a military career was not in LJ's future. Again, we had not pushed that, telling both sons they should determine their own path.

The challenge was getting LJ to campus and settled a couple weeks after I was due in Germany. The solution was Nannette would handle it, complete last-minute actions associated with shutting off utilities and such, then travel to Germany with Karl later. LJ would start his freshman year 3,000 miles and an ocean away from his parents. Fortunately, extended family would support him—Nannette's parents and grandparents were there to help. The family was on the road again, but this time part of it would be left behind. The adventure continued, but at separate locations.

CHAPTER 15

Old World; New World

I HAD OFTEN thought of reporting to a new post as entering a new world. In returning to Germany, I was returning to an old world, figuratively and literally. Along with having been there previously, I thoroughly enjoyed the architecture, customs, traditions, and feel of old-world Europe. But I did not realize that this trip would also bring me into a distinctly and dramatically different new world.

This trip to Germany was not like the last. There was no cargo plane or accompanying family—other than Bandit. Nannette had thought it easier for me to take the dog than her managing that with everything else, including getting LJ settled at Wake Forest. The flight and arrival went well. Picked up by an escort with an Army van at Frankfurt's main airport, I rode the hour and a half to the set of quarters awaiting us. The chief of staff had designated us housing next door to the commanding general in the town of Ansbach, which was twenty-five miles west of Nürnberg in the Bavarian region.

The house was partially furnished with quality European furniture—enough to survive. There was also a pre-shipment of personal items we had sent ahead of our main household goods shipment, which would be enough to get by until the main shipment arrived. The next two weeks were filled with in-processing, meeting the division staff, getting to know the people who would be working directly with and for me, and meeting the various commanders in the 1st Armored Division. I knew some of the commanders from previous encounters, and I was pleased to find two individuals in the division headquarters that I knew well. Lieutenant Colonel Andy Bacevich had occupied a cubicle that backed onto mine in the Department of History at West Point. He was an armor officer who graduated from the Academy a year behind me and previously served with General Franks previously in the 11th Armored Cavalry Regiment. Bacevich was the division G-3, or operations officer, who oversaw the preparation of operational and tactical plans and training for the division. The other individual was my Army-lawyer friend, Lieutenant Colonel John Altenburg. He was the Division Staff Judge Advocate, overseeing legal affairs of the command and the lawyers assigned. It was great having friends and former colleagues around.

Three weeks later, the family arrived. I had arranged some basic things about the house, having spent evenings getting things in order and ready for the family. Bandit wasn't any help but was good company. The real organization began when Nannette was on the ground. We had met General Franks' wife, Denise, in Washington, DC, and she and Nannette quickly bonded in their new roles. Theirs was a friendship that grew strong and deep. As senior ladies for the division and the Command Group, Nannette and Denise had numerous obligations and events to attend, often together. Some of the more interesting Command Group obligations included engaging with the German community. There were many spouse events and activities that they participated in and often led, including conferences, trips, and other projects of the sort Nannette recalled from previous times in Germany. Karl settled in, making new friends and—as always—finding old ones from previous places, like the Altenburg children.

The division chief of staff's duties were varied and broad. They ranged from supervising the staff and ensuring it was operating at peak efficiency, to representing and speaking in the name of the commanding general, both inside and outside the command. An important part of the job was ensuring that the commander's guidance and orders were distributed, understood, and followed. As a relatively new and thus junior colonel who had yet to command a brigade, I took particular care to develop both a professional and personal relationship with the six colonels in charge of the division's major subordinate units, and the others in charge of the smaller specialty units. I approached each individually, explaining that I didn't intend to interfere with their command prerogatives or impose my personal views or approaches. Rather, whenever I spoke it was for the "Old Man," as commanders are fondly called. The frank engagement, followed by action that validated my assertions, gained the other colonels' trust and respect. The CG contributed to this by speaking and acting to show his explicit trust in his chief of staff, whose directives carried the weight of the commanding general.

The broad authority carried other ever-present implications. In later years, I would describe the chief of staff job as one of the hardest, yet most enlightening jobs encountered. There were dozens of small decisions to make every day. The problem was making decisions not as I thought they should be made, but rather as how the CG would desire them to be made. When speaking on behalf of the CG, misrepresenting him or making the wrong decision would be letting him down. That could be a heavy burden when you respect and admire your boss as much as I did. The larger the issue, the more challenging it was.

The first decision was whether to resolve the issue myself or to take it to the commanding general. The question of what to bother the CG about extended to notifications of unusual or consequential events. General Franks instructed me to handle as many of those decisions as possible, as he trusted my judgment. The 3rd Armored Division, like the others, had over 15,000 soldiers. Added to that were probably another 10,000 soldiers of other smaller units residing in the division's geographic area of responsibility and under its general courts-martial authority. With that many soldiers, vehicles, and activities of all sorts, troubling notifications arrived daily around the clock. I usually received an average of three such calls a night, two of which were usually after midnight. After determining the situation and its seriousness, the next decision was whether to wake the Old Man or resolve the issue myself. Often, it was only a notification about something that had already occurred, and nothing more was immediately required. One example involved an inebriated soldier who somehow wandered too close to the West German-East German border. He was apprehended and in U.S. custody before causing a border incident. This could certainly wait until morning, when the soldier's chain of command would need to make a full report. Had he actually crossed the border and engaged with the East German guards, that would have required immediate action. Each morning, I would discuss any overnight actions with the CG. I was a bit surprised and heartened that he never disagreed with my resolution of the issues. It was good practice for higher command—and an indispensable lesson on how to trust your subordinates.

I found General Franks to be not only an excellent tactical and operational commander and leader, but also a great teacher, mentor, and friend. I watched my boss pass on wisdom gained through years of experience and how he encouraged growth in subordinates. I learned how to exert leadership across a large and complex organization where it was not possible to physically engage everyone, while ensuring your intent was understood and the command climate you desired was implemented. Perhaps one of the skills I admired and sought to emulate most was the ability to know just when to make a decision. Unfortunately, some commanders do not know when to do so. They might make a quick but hasty decision either before having all relevant information or before a decision is necessary or prudent—or they may wait too late for the decision to be effective or even relevant. I also valued how General Franks analyzed the various consequences of actions and decisions. This applied to tactics on the battlefield as well as discussions on logistics, personnel management, or policy. Every course of action has pros and cons. The commander's job is to sort them and choose the best overall course of action. The thing that separates good commanders from the rest is how quickly and decisively they can do all that, when the pressure is on.

The senior leaders of the division formed a high-performing, well-coordinated team that made Old Ironsides a remarkable fighting machine. There were four U.S. divisions in Germany, along with West German Army, the Bundeswehr, and British divisions. Within NATO, there were additional divisions from the Netherlands and Canada, plus other smaller units. The French also had divisions which could be counted on to defend Western Europe. But I and many others objectively believed that none was better trained and prepared than Old Ironsides.

As with all units in the Army, the team was constantly losing and gaining members as assignment times, enlistments, schooling, or other events caused personnel changes. Though it felt far too soon, the division leadership began to turn over. Short of the two-year point, Major General Franks was promoted to lieutenant general and selected to command VII Corps, which consisted of two divisions. I received orders to report to the 3rd Infantry Division to command a brigade in Kitzingen, Germany. As 3rd Infantry Division was part of VII Corps, I would continue working indirectly for my old boss.

The 3rd Infantry division was one of the oldest and most decorated divisions in the Army. I felt honored to have served in four of those highly respected divisions. The 1st Infantry Division, The Big Red One, was the oldest, and had served in World War I, World War II, and Vietnam. The 1st Armored Division and the 2nd Armored Division had fought hard in North Africa and Europe in WWII. The 3rd Infantry Division—which I was now joining—was nicknamed "Rock of the Marne" for its gallantry during World War I, fighting along the Marne in France. It had been the unit of Audie Murphy, the most decorated soldier in World War II. In fact, one of the battalions I would command in the brigade—the 1st Battalion, 15th Infantry—was Audie Murphy's old unit.

Arriving in the division having already served as a division chief of staff probably led some of the other colonels to expect me to assume a superior or disdainful attitude. If so, they were disappointed. Rather, I sought the same sense of teamwork I had left behind at 1st Armored Division. I quickly found myself deeply involved in training and focusing on those things that had been important when commanding a battalion. One lesson I had learned was that in a battalion, you could recognize nearly every soldier in your unit by sight if not by name. In a 4,000-man brigade, you could not. Rather than commanding those soldiers directly, you commanded three or four competent and independent-minded battalion commanders. It was my job to get them to work together as a team.

Commanding the brigade was enjoyable and rewarding, but not without challenges. I needed to allay the natural concerns my new division commander, Major General Shoffner, might have about my being a loyal team player. After all, he was now working for General Franks, whom I had just served intimately as his chief of staff. While Franks and I enjoyed a professional friendship, neither ever gave the division commander cause for concern, and I gained the commander's confidence and admiration. Still, I felt that being a friend and former associate of your boss's boss wasn't all bad.

There was an incident early in my command that could have caused concern if not handled quickly and properly. One of the three maneuver battalion commanders had already been in command for eighteen months and was thought of as one of the best and most aggressive in the division. I observed and agreed that he appeared, as was said, to "run a tight ship." Still, that commander seemed to exude a bit too much self-assurance. That did not particularly bother me, as I thought commanders—especially armor and cavalry leaders—needed some of those traits. It wasn't until the brigade legal advisor, a young JAG officer, suggested that I review the entire brigade's statistics for the use of non-judicial punishment that a serious concern arose. Commanders could administer non-judicial punishment, short of a formal court martial, for minor infractions. Looking across the results of each unit for the previous year, the battalion commander in question was different from the others in a few ways. He had administered more punishment than the others, in keeping with running a tight ship. On average, his punishments for specific offenses were high, but not out of bounds. They also appeared equable across all soldiers punished that way. But in some cases, some or part of the punishment was suspended and set aside for a period of time while the soldier demonstrated corrected behavior. At the end of the suspension period, the punishment was completely forgiven if no further misconduct was observed. I had used this technique myself and thought it good. The only problem, upon careful examination, was that the battalion commander in question invariably suspended sentences and punishment for most Caucasian soldiers but not minorities, especially African Americans. The latter, in every case, received full and immediate punishment.

After considering the circumstances and how detrimental such disparate action could be to morale and good order and discipline, I felt compelled to act decisively. The question was how to help that battalion commander and others to be introspective about their leadership and any bias, subconscious or otherwise. I held regular weekly meetings with the commanders and staff to discuss upcoming events and receive feedback; I asked the JAG officer to distribute the non-judicial summaries to the respective commanders at the next meeting. The JAG maintained those summaries routinely, and commanders were encouraged to review them periodically. I asked them to do so then. A week later, I directed the battalion commander in question to come to my office. When asked if he noticed anything unusual in his report, the lieutenant colonel replied that he did not. When I pointed out the racial disparities in suspended punishments, the officer reddened, then exclaimed that he was not racist. I explained that I wasn't alleging he was but wanted to know if he had an explanation or been aware of differences in administering punishment. Completely abandoning his usual abundance of poise and self-assuredness, the lieutenant colonel emotionally expressed denial of any bias. I opined that the report would show otherwise and instructed him to return to his battalion and carefully consider whether he had subconsciously or consciously treated soldiers of color differently. I added that such actions, whether actual or merely perceived by his superiors, could seriously impact his career. Badly shaken, he left the office.

Subsequent reviews of the lieutenant colonel's judicial actions revealed none of the previous bias. I was certain he had learned a very hard lesson. However, I was never able to convince myself that the officer did not hold deep seated prejudices, whether conscious or otherwise. The evaluation report I prepared when the officer left command acknowledged his superior tactical performance, though it also subtly indicated the possible need for continued growth in certain leadership and human relations areas. It was far from a career-killing report but would alert others and the officer to be on guard. The separate report on non-judicial summaries that the JAG submitted through his technical chain may have flagged the officer in question within other circles, as well. Years later, I felt justified in my actions upon learning the officer was never promoted nor in command of soldiers again. For me, there was no place in the Army for prejudice or leaders who were not evenhanded in caring for soldiers. Obviously, others had the same concerns about him.

A highlight from my time as a brigade commander was the extremely improbable fact that one of the other two maneuver brigade commanders was my close friend Colonel Larry Ellis. There were roughly thirty maneuver brigades in the entire Army that infantry and armor colonels could command. Of the thirty, perhaps three or four were commanded by African Americans at any given time. The probability of us being assigned together was beyond calculation, but it had happened—just like having two classmates in the same company back in Vietnam. This time, it truly was by happenstance.

While I had spent my first year in Germany as Old Ironsides' chief of staff, Ellis had commanded the 1st Brigade, 3rd Infantry Division, in Schweinfurt. When I assumed command of the 2nd Brigade in Kitzingen, located twenty miles to the south, two great friends were reunited. Socially and professionally, we enjoyed being together, as did our families. Many who did not know us or our past relationship well assumed that two Black brigade commanders in the same division—which was unheard of given the relative few that had the command opportunity—would be natural adversaries, competing for recognition. They did not know the extent to which we each admired, respected, and liked the other.

That expectation of rivalry extended to maneuver training at Hohenfels Training Area. Much like at Fort Hood, brigades in Germany sometimes had one battalion play the OPFOR role. Upon observing the brigade's tactical performance, the division commander assigned 2nd Brigade the mission of playing OPFOR for the battalions in the other two brigades. I enjoyed the mission, although it meant spending three times longer than the other brigades at the training area. I considered it the cost of being fully proficient and assigned the important role. Plus, it was additional maneuver time, which you could never get too much of. Still, the soldiers and I missed our families and the comforts of garrison.

The training called for each battalion to confront a Soviet Motorized Rifle Regiment, or MRR, of about seventy-five infantry vehicles and perhaps thirty tanks. My brigade had about forty-five infantry vehicles and 116 tanks. I could not replicate a Soviet MRR. I could, however, easily replicate a Soviet Tank Regiment of ninety-two tanks. The problem was a tank regiment had far more firepower than a Rifle Regiment. The result of that and our high level of tactical proficiency was that the simulated battles were usually as one-sided for Blue Forces as most had been at NTC. I took delight in attacking the battalions of Colonel Ellis, or "Raider 6" as he was known on the radio net. I, known as "Spartan 6," would faithfully employ Soviet doctrine and tactics, which Ellis and I had both studied to better fight against them. Although the superior combat power from ninety-two tanks usually resulted in lopsided battles, that was exactly how commanders and their units engaged in the training's hard lessons.

The general misperception about Ellis's and my relationship among many of our subordinates fueled one thing we would laugh about for many years to come. Many evenings, after concluding the day's battle and exhaustive AAR, we two brigade commanders were seen in animated discussion in the Training Area's cantonment. Subordinates from either brigade avoided getting too close, assuming we were butting heads about what had occurred, which heightened the perceptions of rivalry. Ellis and I were actually comparing notes to help improve the battalion being evaluated and joking about what had happened. Sometimes, I would ask Ellis what he wanted done to a particular battalion. He might reply that a certain commander had short changed defensive training and "needed a lesson." To that, I would reply, "I'll have something special for him tomorrow." There was certainly rivalry, but of a friendly type most outsiders only recognized after knowing us both well.

At the end of my first year in command, Ellis completed his second. His change of command involved a move back to the Pentagon. We said farewell to our friends, sad to see them go, but pleased that the Ellis and Lumpkin families would be near each other again. Harry Lumpkin had changed jobs but was still in the Pentagon. Before parting, Ellis and I had talked about the future. Both of us were approaching twenty-three years of service. Upon reaching the twenty-year point, we had been eligible for retirement but were still having fun. One more assignment would probably bring us to the twenty-six-year mark, which was when many colonels retired. That maximized retired pay for the time served, and left sufficient work years for a second career. We decided to hold on and see what the future brought.

The Jordan family continued to enjoy the area, frequenting favorite *gasthauses*, small German neighborhood restaurants, often family owned. Karl had learned to drive, obtaining his stateside driver's license while visiting his grandparents one summer. He had yet to take the dreaded German driving exam but would do so soon. The small housing area was home to all the battalion commanders in the brigade and most of the staff officers. Part of the area was a small officers club that had been a Luftwaffe club during World War II. In fact, the prototype German fighter jets had been based in Kitzingen. The O Club was the center of community activities for the officers and offered good meals and excellent beverages. Every Wednesday, there was a family night special with special menus. Neighborhood teens often went to the O Club for dinner that evening, the friendly staff charging the cost to their parent's account. Adults would go later and enjoy adult time together, the spouses freed from kitchen duties that day.

LJ continued at Wake Forest and was doing well, though the family regretted not being able to visit him regularly or see him play football. We saw two games in his four years. We happened to be back in the States for a few days for the first, where Wake Forest played West Point at the Academy. Friends in the stands asked who I was rooting for. "Army, naturally," I responded, but did cheer every time my son made a tackle or good play. The second was at Wake Forest, where I was surprised to discover that Wake Forest's home uniforms were almost identical to the Army team's. But the real surprise came at the end of the game when the school's alma mater was played. The tune was the exact same as West Point's. Quite a double coincidence. Notwithstanding our infrequent trips to the States, the family did manage to keep up with LJ through weekly calls. Sometimes, however, we would receive calls from an excited or dejected LJ—depending on the outcome of the game—late at night or very early in the morning. With forced patience, we would remind him of the six-hour time difference between North Carolina and Germany.

▼ ▼ ▼

The second year in command passed much like the first, to a point. I knew that a board was being convened in Washington, DC, to consider colonels for promotion to brigadier general. While I was beginning to think that I might have a slight chance of being selected someday, I did not seriously think it would occur until after completing command and receiving an Officer Evaluation Report, or OER, recording my final and overall evaluation. Then one night, the doorbell rang. Opening the door, I found no one there, but on the porch was a bottle of champaign with a card that read "Congratulations." It was from the brigadier who lived two doors away and served as one of the Assistant Division Commandeers, or ADC. When encountering the ADC the next day, I thanked him for the bottle and asked if it was for some good news. The ADC merely winked and smiled. Army promotion lists were distributed early to senior commanders at the general officer level. All were forbidden from disclosing the names prior to the appointed date and time for public release. There were severe consequences for failure to comply. Convinced that the improbable was about to occur, we quietly celebrated in private, awaiting the public announcement. It came two days later. Interestingly, that ADC was the same brigadier I had encountered in the SECDEF's office while trying to be released from the Joint Staff to go to Germany. Again, it reminded me that prior encounters and good reputation helped in unexpected ways.

As a promotable colonel, I was about to enter a new world. The mystique associated with generals often overshadows the truth—being a general was mostly very hard work. The public saw the pomp, the ceremony, the small silver stars of rank, and the people assigned to assist, like aide-de-camps. What the public didn't see was the often-crushing responsibilities, multitude of tasks, ceaseless demand for selfless service, and expectations of both superiors and subordinates for superhuman performance. It would be a tremendous honor to be a general. It would also be a tremendous weight to bear.

Meanwhile, I still had a brigade to command. Another momentous event occurred that autumn. Saddam Hussein invaded Kuwait. Operation Desert Shield began. VII Corps was sent to Saudi Arabia to team with XVII Airborne Corps from Fort Bragg to assemble the force that would prosecute Operation Desert Storm in the combat to follow. Prior to the rise of tensions in the region, the Berlin Wall had fallen, and the Soviet Union began to dissolve into Russia and a few former Soviet republics that became independent—at least in name. Resultingly, the Army had begun to drawdown as part of the nation's peace dividend. One of my units, the infantry battalion, had begun to turn in equipment and lost half of its manpower. When units were selected to go to the desert, the brigade's officers had been certain we would go. The brigade had a superb reputation for tactical mastery; General Franks would certainly want his old chief of staff leading a brigade in the fight to come; I had been selected for Brigadier General. The officers reasoned that everything had aligned. The problem was we did not have a combat effective infantry battalion. Another brigade that had not begun the drawdown process was selected to go instead. The officers were emotionally crushed at missing the opportunity to go into combat as they had all trained to do. Very few were old enough to have served in Vietnam, and the vast majority had not participated in any of the smaller operations like Operation Just Cause in Panama. As Operation Desert Storm went underway, I received orders to give up command of the brigade, fly to Saudi Arabia, and assume the position of assistant division commander in the 3rd Armored Division, nicknamed "Spearhead." Although saddened they were not to go, the brigade celebrated their commander being elevated. We had a quickly organized farewell at the Officers' Club, where I first uttered the lines about a special place, a special time, and a special group of people. The 2nd Brigade, 3rd Infantry Division, in Kitzingen, Germany, was certainly all three.

Meanwhile, the German community rallied behind the Americans departing for the Gulf and those remaining behind. Both Americans and Germans understood the security needs, as terrorist groups had attacked their facilities in the past. During our first trip to Heidelberg, Germany, the Baader-Meinhof Gang had attacked the U.S. Army commander in Europe's armored sedan with rocket propelled anti-tank rounds. He survived due to the reinforced and armored "hard car," but the attack rattled the country. The Baader-Meinhof Gang continued their attacks against German politicians and others, or at least until international law enforcement was able to silence the group. In response to the threat of Islamic extremists' potential reactions to Desert Shield and Desert Storm, the U.S. Army posted Abrams tanks and Bradley Infantry Vehicles at the entrance to the housing area, and the German government took solid steps to assist. The County Commissioner of the Kitzingen area was extremely helpful in particular. The security actions they took greatly comforted the surrounding families, including my own. I was also comforted by the fact that community relations were extremely cordial. I had developed a genuine friendship with local civilian officials and leaders, who truly made the Americans feel part of the community.

CHAPTER 16

Brigadier General Stars

Arriving shortly after the one-hundred-hour combat operation had ended, I joined the 3rd Armored Division. Though offensive operations had ceased, the situation remained volatile and uncertain. There were concerns that rogue Iraqi elements might launch sneak attacks or initiate an insurgency. The area was littered with mines and unexploded ordnance, and reports of leaking chemical munitions and other contraband were mounting. Oil well fires continued to burn, casting toxic fumes and a hellish glow at night.

While most other units had departed, the 3rd Armored Division stayed behind to ensure stability and secure excess supplies and the thousands of pieces of U.S. equipment brought in. Surveying the desert landscape, I marveled that nations would fight over it—were it not for the oil buried beneath its sands. One day, as the temperature climbed to 125 Fahrenheit and the strong wind gusts felt like a blowtorch, I wryly wondered whether the Good Lord had called it a day before finishing this part of the world.

Eventually, my orders arrived authorizing frocking, allowing me to wear the star of a brigadier general. For the first time since Vietnam, I had my new insignia pinned on—this time, without Nannette present.

As other units returned home, we became the last remaining American division in Iraq. In May 1992, the long-awaited order for redeployment finally came. The 1st Brigade was scheduled to depart first, and flight manifests were prepared. But just as they were preparing to move to the airfield, word came down: The departure was postponed. The brigade would remain until the 11th Armored Cavalry Regiment arrived from Germany to assume the mission and maintain the U.S. presence.

Understandably, disappointment ran deep. Families had already been notified that their loved ones were coming home. Major General Jerry Rutherford, the division commander, regretted having to break the news. I volunteered to deliver the order. The officer in command of the brigade, Colonel Bill Nash, was a fellow West Point classmate. He and I shared a bond because he, like me, was recently promoted to brigadier general and he was a brother armor officer. We were the first two members of our class to reach the general officer ranks. Even so, that did little to ease the burden of relaying such agonizing news. Still, I passed the order.

Once the 11th ACR had arrived and assumed responsibility for the sector, we were finally cleared to return home. By then, another longtime friend, Colonel Andy Bacevich, whom I had served with at both West Point and in Old Ironsides, had taken command of that regiment. Once again, my path intersected with those from earlier times in my life.

Social activities tied to the unit resumed after our return from the Gulf. Two ceremonies in particular stand out—both commemorating the victory in Operation Desert Storm.

At the first event, Nannette and I represented the United States at a ceremony hosted by the French Army. It was a rare gathering involving that particular ally and our first time witnessing the French Foreign Legion in formation. Dressed in their distinctive uniforms, the Legionnaires marched in their signature slow, rhythmic cadence—a striking contrast to other military styles.

We were seated just one row behind the British representatives: Prince Charles and Princess Diana. Sharing space with royalty was a first for us and made for quite the story. Reflecting on the moment later, I was reminded of something someone once told to me, "No matter where you're born or what circumstances you're from, you can rise. With hard work, perseverance, strong values, and the support of the Lord, you can stand humbly but confidently among the powerful, the famous, and the acclaimed."

The second event took place in Stuttgart, Germany, commemorating VII Corps' participation in the Gulf War and marking its deactivation as part of the postwar drawdown. Much of the Army senior leadership was in attendance, and the ceremonies concluded with a large formal reception.

The guest of honor was General Colin Powell, Chairman of the Joint Chiefs of Staff. As he moved through the room, Powell's extraordinary recall of names stood out. I had met him only briefly months earlier at the General Officers Orientation Course in Washington, DC—known informally as the "Charm Course." There had been about forty newly minted brigadiers in attendance, and the Chairman likely participated in such events monthly. Yet, as he approached me, he asked, "How are things going, Larry?" Caught off guard, I barely managed a response.

Later, his aide passed by and joked, "Don't get a big head. The Boss just remembers names." Still, being remembered by the Chairman of the Joint Chiefs of Staff felt like a true honor.

Spouses generally enjoyed attending such events, as long as they weren't drawn into the formal agenda, because it gave them a chance to socialize and reconnect with friends. They much preferred that to being grouped with their husbands during long, often tedious professional discussions. This event was no exception. Nannette and the division commander's wife, Betty, had developed a warm friendship based on shared experiences, and their conversations flowed easily between lighthearted jokes and more serious reflections.

As clusters of general officers became absorbed in hearty conversation, the two women, along with a few other spouses, gradually moved to the periphery of the room. After a few minutes, unnoticed by their husbands, Betty turned to Nannette and quietly said, "They also serve, who only stand and wait." Both exchanged knowing smiles. The remark, poignant and true, captured the often-unseen strength of military spouses—those steadfast presences who make family life possible amid the demands of service.

As one popular T-shirt sometimes worn by Army spouses reads, "Army wife—hardest job in the Army."

The Army formally recognized the vital role spouses play by adopting the term "Command Team" to describe the partnership between commanders and their spouses in supporting unit and family welfare. More than any other branch, the Army is a married force, with a higher percentage of married service members than the others. It has long understood that while it recruits soldiers, it must retain families. If families don't feel welcomed and supported, soldiers are unlikely to stay in.

To that end, Family Support Groups, which were organized and run by spouses with backing from the chain of command, were officially acknowledged and resourced. Spouses were also invited to attend pre-command courses for incoming battalion and brigade commanders, recognizing their essential leadership roles. Much of what was taught in those courses was already familiar to Nannette, who had grown up as an Army brat and observed her mother and other senior spouses in action. She refined that knowledge through experience. Nannette served as the senior spouse during my company, battalion, and brigade command tour, having learned through the mentorship of more senior spouses who generously shared their wisdom and friendship.

Army spouses are, in many ways, a sisterhood—deeply committed and quietly capable. Most are as dedicated to the mission as their soldier partners and remain humble in their roles. Few seek to "wear their husband's rank," and those who do often find themselves without close or lasting friendships.

The year we remained in Germany was eventful. Training operations resumed, and our engagement with the German community intensified; we often participated in the community through both official business and social, or public relations, events. German political attention was focused on questions about future U.S. troop levels, the repurposing of now redundant American military facilities, and the integration of former East Germany.

The apparent end of the Cold War and the reunification of the two German states brought great change. Travel to Berlin, once bureaucratically burdensome, became simple. The division commander arranged a trip to the city for unit leaders. Having visited Berlin several times, we already considered it one of our favorite places, but traveling freely through what had once been East Berlin was a novel and moving experience. Few visible traces of division remained—aside from the noticeably older, more worn appearance of much of the eastern sector.

One memory stands out vividly. On the weekend following the first free elections in the newly reunified Federal Republic of Germany, we stood at the Brandenburg Gate. Once a symbol of division, the gate had straddled the line between East and West, with no passage permitted along the broad boulevard beneath it. But on that day, German citizens strolled, biked, and drove freely through it. It was a powerful symbol of change, and it was a moment we never expected to witness in our lifetimes.

There had been changes in the family as well. While I was deployed to Desert Storm, Karl had participated in a student exchange program in the Czech Republic, briefly living with a host family. He enjoyed the experience immensely, and it broadened his awareness and perspective, helping to cultivate a more global outlook.

▼▼▼

Our four years in Germany were memorable in many ways. During our last Christmas there, LJ flew over for the holidays as he had in previous years. It was his senior year at Wake Forest, and he asked if he could bring along his girlfriend, Erika, a junior at the same university. We had met her the year prior on leave in the States. LJ and Erika had recently become engaged.

Fascinated by Germany, Erika delighted in the sights and holiday traditions, especially the spectacular festivities of the German Christmas season. The highlight, however, came when we were told we'd be attending holiday cocktails with a German princess in her castle. She thought we were joking—until we told her to dress for the occasion. The event was a part of a series of community relations functions I attended in my role as Assistant Commander and Senior U.S. Representative in the area.

By the end of the visit, Erika remarked that perhaps LJ should "join the Army like his dad." We explained that the life of a brigadier general was quite different from that of a newly commissioned lieutenant. The spacious home we now occupied was quite different from the stairwell apartments we had previously occupied during earlier assignments, and official dinners weren't exactly how junior officers, like LJ, would start out. It had taken over twenty-three years—and even that was quicker than most—to reach that point.

I dismissed the idea, as LJ had expressed no real interest in a military career since his earlier curiosity about West Point. He had even declined to enroll in ROTC at Wake Forest, saying that football took up all his free time—and that, after spending the first eighteen years of his life moving with an Army family, he'd already done his time.

Karl had become a fine football player in his own right, also playing a defensive back like his brother. Over four years, he attended three different high schools—Ansbach, Würzburg, and Hanau. He ultimately decided to apply to West Point. We supported his decision, though, as with his brother, we did not push him to do so.

It seemed another soldier in the family and another generation might be serving the nation—until, the Friday before his scheduled medical and physical fitness exams, Karl suffered an ACL injury to his knee. The injury required surgery and left him on crutches for a couple of months. His doctors informed him that competitive sports might no longer be possible, and as a result, he was no longer physically qualified for West Point.

After graduating from high school, Karl enrolled at Syracuse University. But his football story didn't end there. A year later, after rigorous rehabilitation, he tried out as a walk-on and made the Syracuse football team. Walk-ons don't receive scholarships, which are strictly limited by the National Collegiate Athletic Association (NCAA) regulations, but Karl played for two seasons.

In his senior year, he surprised us by stepping away from football to concentrate on achieving the highest grade point average possible. Years later, during a conversation about that time, he confided that his goal had been simply to prove to himself that he could make the team. I told him I was prouder of that than if he'd earned a scholarship and played all four years.

Karl would not become a soldier, but he had the heart and spirit of a warrior.

The last notable incident from our time in Germany involved a position that I didn't get. The General Officer Management Office (GOMO), which handles assignments for generals—much like branch officers do for colonels and below—had informed me that I would be assigned to Fort Knox as Assistant Commandant of the Armor School. It was an exciting opportunity, as the Assistant Commandant essentially ran the school, while the formal commandant role belonged to the Armor Center's two-star commander. I looked forward to taking on the task.

Then, unexpectedly, I received official notice to fly to Washington, DC, to report to an office in the Pentagon for a meeting. No further information was provided, except that it was a high priority matter. Upon arrival, I noted that the meeting location was on the E Ring—the corridor where senior military and DoD leadership offices are located.

What followed were three quick interviews with different individuals I did not know who asked a range of questions. At the end, I was told I would be contacted later. Still puzzled, I returned to Germany.

Two weeks later, another mysterious order arrived—this time to report to NATO Headquarters in Brussels. There, I met Pete Williams, the Department of Defense Press Secretary, whom I recognized from televised briefings. After about twenty pleasant, but puzzling, minutes of conversation, he said, "It's time to meet the Man."

When asked who that was, he simply said, "SECDEF."

The next meeting was with the Secretary of Defense Dick Cheney, who was relaxed and direct. He explained that he was looking for a new executive assistant to replace the departing naval officer currently in the role—skipping over two-star entirely. The implication was clear: serving directly under the Secretary of Defense could be a significant career accelerant.

The SECDEF explained that timing was critical. He was about to begin a major trip to the Pacific region and needed a replacement immediately. I responded that I was honored to be considered—but couldn't leave Germany at once.

When he asked why, I explained that I had been stationed there for four years. My younger son had graduated three different high schools and was graduating in a month. My older son was graduating from college. My family would need help relocating from Germany back to the United States.

The SECDEF replied, "Someone can probably help them."

I was a dedicated officer who followed orders but wasn't prepared to leave my family in such a lurch in a foreign country for the sake of a job. I also wasn't certain I wanted to work in an environment where leaders didn't seem to understand the demands placed on military families.

When he asked me what I truly wanted to do, I answered simply: go to Fort Knox and help train and lead the Armored Force.

He thanked me and the meeting ended.

Afterward, several people told me, "You just can't turn down a job offer from the SECDEF." They said I had thrown away the best opportunity of my career.

Perhaps I had. But I felt more than ever that I had made the right decision. It reminded me of a moment twenty-two years earlier in Vietnam, when I chose not to extend my tour just to take a so-called plum assignment. In both cases, I put my family ahead of my career.

Something close friends and I had often repeated came to mind: "Manage your own career—or someone else will mismanage it for you."

Within a few weeks, we began the familiar process of packing once more—this time headed for our new home: Fort Knox.

CHAPTER 17

Finding Gold

To most Americans, the mention of Fort Knox evokes the U.S. Department of the Treasury's Bullion Depository, which is a gold vault. Few people have ever seen inside that most secure facility—and I hadn't realized, upon arriving, that I would be one of them. That massive cache of bullion was certainly impressive, but it wasn't the greatest treasure I would discover at Fort Knox. As I prepared to arrive, I was uncertain what kind of reception to expect.

After twenty-four years as an armor officer, I returned to Fort Knox, the Home of Armor, for only the third time. My first had been as a West Point cadet during summer training for rising Third Classmen (sophomores) when we toured various combat branch schools. That experience was when I first felt certain I wanted to pursue a career in armor. My second visit was as a new lieutenant attending the Armor Officer Basic Course. Although I had made several brief visits since then, unlike many tankers and cavalrymen, I had never been stationed there.

In contrast, the two-star general commanding the Armor Center had deep roots at Fort Knox. He had served there twice in addition to completing both the Basic and Captains Advance Courses and was well-acquainted with many of the long-time civilian employees and community in the surrounding civilian communities. I, on the other hand, would need to build those relationships and gain their acceptance.

There was some trepidation on my part, as no African American general officer had ever been assigned to the Armor Center. I also recalled an unpleasant social experience during my cadet orientation trip. Each installation we visited hosted social events during our free time, part of the Branch Schools' efforts to impress the cadets and attract future lieutenants. Most arranged for young women from the local community to attend, promising a chance to meet talented—and supposedly charming—young men. Care was taken to pair each cadet with a young woman by name.

This evidently was even more deliberate for the Class of 1968, more minority cadets than any class before it. But the carefully choregraphed plan unraveled, and the pairings ended up being assigned at random. The result was a number of shocked, dismayed, and—in some cases—outraged young women who hadn't expected to be paired with African American, Asian, Hispanic, or Jewish cadets. The incident left an indelible mark on those of us who were abhorred by their assigned dates for the evening.

A few of my white classmates offered vague apologies, saying, "That's just the South." I couldn't help but wonder whether nearly a quarter of a century had changed anything.

As it turned out, the community was both welcoming and far more socially enlightened than I had feared based on my earlier experience. Running the Armor School proved to be both exciting and rewarding, thanks to the breadth of responsibilities involved. The school conducted courses for lieutenants and captains, as well as junior and mid-grade noncommissioned officers (NCOs). These courses were mandatory for all armor and cavalry officers and NCOs in the force.

In addition, the school hosted the Master Gunner Course, which was a graduate-level program designed to produce gunnery experts who would serve as indispensable resources for battalion commanders during gunnery training. The officers and NCOs teaching at the Armor School were among the best I had seen. They took pride in, and found joy in, producing well-trained mounted warriors.

One instructor stood out in particular: a master sergeant serving as the NCO Academy commandant, who was doing an amazing job. Years later, I would serve with him again in Europe. That great NCO would go on to become the Sergeant Major of the Army, serving as the senior enlisted advisor to the Chief of Staff and the highest-ranking NCO in the Army. Sergeant Major of the Army Ken Preston was one of the long line of excellent NCOs I had the privilege to serve with, admire, and learn from.

My move to the Armor School also reconnected me with my former boss and mentor, General Fred Franks, who was now a four-star general and the commander of the U.S. Army Training and Doctrine Command (TRADOC). This command oversaw all the Army's branch schools and centers under its command. It was also during this time that I came into closer contact with another very senior officer, the Chief of Staff of the Army, General Gordon Sullivan.

Although we had crossed paths briefly at a few large functions, our connection went further back. When I transferred from the Big Red One in Vietnam to a separate cavalry squadron to complete my tour, then-Major Sullivan was the assignment officer who directed me to my new unit. Years later, when I served as Chief of Staff for General Franks, Sullivan—now a division commander—led the 1st Infantry Division and participated in a REFORGER exercise. We met briefly at an AAR.

Not long after I arrived at Fort Knox, I received a call from General Sullivan. "I had that job you are in," he said. "Don't screw it up!" I was momentarily caught off guard but quickly realized he was joking—after all, he was the one who had selected me for the job.

Then he got serious. He gave me a clear and unexpected task: determine why there were so few minority officers in Armor branch—and fix it.

The Army had been working toward equal opportunity and treatment of soldiers since President Truman's Executive Order 9981 in 1948, which mandated desegregation of the Armed Forces. That is not to say the effort was always wholehearted effort or free from resistance by individuals entrenched in prejudiced views. By 1992, however, the Army had implemented a number of initiatives, especially during and after the Civil Rights Movement, to address racial disparities.

Despite this progress, certain branches continued to show disproportionately low numbers of minority officers, with armor and cavalry among those most affected. I was directed to take the time necessary to analyze the root causes of this disparity and to devise a strategy for addressing it.

It would take more than a year to complete the work and begin seeing initial gains. First, I had to think through the issue and decide how to attack it. Having served as a minority officer and navigated obstacles—whether real or perceived, overt or subvert, intentional or unintentional—I believed I understood some of the underlying causes of the disparity. But as I delved deeper into the study, I came to understand much more.

I spent time discussing the matter with the Army's senior personnel officer, Lieutenant General Ted Stroup. I also decided to interview several dozen active-duty officers—both minority and majority—across a range of ranks. My goal was to gauge their perceptions of the Armor Branch, the overall quality of its officers, and how minorities were viewed and managed within it.

Equally important, I interviewed several retired African American generals and colonels to gain their long-term impressions—what obstacles they had faced during their careers, what factors influenced their decision to remain in the branch, and what had enabled their advancement in the Army.

My research uncovered many interesting—and, in some cases, surprising—findings. With the assistance of the Army Personnel Officer and his staff, I learned that over 40 percent of all armor and cavalry officers came from just five institutions: West Point, Virginia Military Institute, Texas A&M, The Citadel, and Norwich. All are fine schools that have produced outstanding Army officers. However, none had historically commissioned many officers of color.

While African Americans, Hispanics, Asians, and other minorities were beginning to graduate from West Point in more proportionate numbers, the other institutions still had further to go. Though this concentration was the result of historical happenstance rather than intentional exclusion, the fact that such a large share of branch officers were being drawn from so few institutions meant that many talented individuals were likely being overlooked. It also meant achieving a more diverse group of officer corps would not happen without deliberate effort.

The Army—and the Armor Branch in particular—needed its officer corps to reflect the diversity of the nation it served. It could not afford to pass over talent.

I set out to develop a strategy to address the low number of minority accessions among incoming lieutenants. This involved efforts to attract both Army and ROTC cadets to the Armor Branch by highlighting the opportunities it offered and its position at the cutting edge of modern warfare technology. I also stressed that many of the Army's senior leaders come from the combat arms—armor, infantry, and artillery.

In addition, I worked to influence school administrators who often, perhaps unknowingly, steered minority ROTC students away from the combat arms. They did not always recognize the leadership and management experience gained in these branches was just as valuable as the technical skills supposedly found in the so-called technical branches.

I also examined ways to improve the retention and advancement of qualified minority officers. One key factor that surfaced early was the critical role and importance of mentorship for junior officers. Another was the perception that minority officers had fewer opportunities to secure broadening assignments—such as scout platoon leader roles, Training Board positions, and West Point instructor posts—assignments I had been fortunate to receive, and which were instrumental in my own career development.

It took two more years before I was able to show substantial gains in minority accessions, bringing the percentage of minority officers entering the Armor Branch in line with that of the Army as a whole. Improving retention rates, however, would require sustained effort across the force—through both policy and practice.

I presented the findings of my study and the results of our efforts to the Army Chief of Staff. He was impressed by the progress and asked me to share the study with my fellow school commandants, especially those leading the other combat branches. I was a bit surprised by their responses. Some felt their branch had faced more pressing challenges, or that minority accessions weren't a problem for them in the first place. However, two commandants took the study seriously and began implementing their own comparable strategies.

Interestingly, both of those individuals went on to be promoted to lieutenant general while most of the others did not. Whether or not that was the reason, I concluded the Army rewards leaders who demonstrate initiative and a commitment to helping every soldier reach their full potential.

▼ ▼ ▼

While the study on minority offers was important, it was not my main focus. As the Assistant Commandant, I oversaw the daily operations of the Armor School and the many training and study courses it conducted. With so many activities underway, time passed quickly.

At the same time, the Army Chief of Staff sought to use technology to empower the force more effectively—a focus that would soon become central to the mission of the Armor Center. Meanwhile, the Armor

School continued to serve as the primary training institution not only for Army tankers, scouts, and vehicle mechanics but also for members of the U.S. Marine Corps. Given that the Marines Corps maintained only three active and one reserve tank battalions—compared to over fifty battalions and squadrons in the active Army and additional units in the National Guard—it made sense for them to utilize Army training rather than replicate it themselves.

In addition to the U.S. forces, allied students from around the world were also enrolled. Altogether, the Armor School trained more than 20,000 students annually. It also developed and published tactical and training doctrine for armor and cavalry forces. Both my family and I settled into a busy but enjoyable routine.

It was a memorable time for our family. LJ graduated from Wake Forest shortly after we returned from Germany and entered the job market. Fort Knox was only about two hours from Lexington, where Nannette's parents lived, making it easy for the family to meet in either location. Her parents enjoyed visiting Fort Knox and being back on a military base, surrounded by the familiar sights and routines. Top Owens especially liked reminiscing about completing his Basic Training there in 1950 and reflecting on how much had changed since then.

That first autumn Karl had just begun his studies at Syracuse. Being back in the U.S. also meant we were able to spend more time with extended family. My sister Joyce and her husband came to visit, bringing their precocious and energetic eight-year-old son, John. Fascinated by the tanks and all he saw at Fort Knox, John was full of questions and curiosity. At the time, no one could have predicted where that fascination would lead—but it planted a seed. Years later, John would enter the Army after college as an armor officer, adding to a family legacy of Army service that I was still only beginning to appreciate.

The most momentous family event was LJ's marriage to Erika, which followed her graduation from Wake Forest. The newlyweds launched a life together far removed from the military. In time, both became high school teachers—continuing the longstanding family tradition of service through education.

Home life for the Jordans changed with Karl's departure for college. For the first time since my return from Vietnam, it was usually just the two of us. We joked about enjoying the freedom and only missing the "child labor" that had once helped with mowing the lawn and other tasks. Still, the house felt strangely quiet. Even Bandit seemed to miss the extra activity that two boys had brought to our home.

Nevertheless, we enjoyed life at Fort Knox. After living in Germany, where it was different from being on a stateside military installation, and in civilian communities like Northern Virginia, and during graduate school, the routines and atmosphere of a U.S. post felt familiar and natural. Each day began with reveille—a single cannon shot, a bugle call, and the raising of the American on the tall flagpole by the parade field. The same ceremony closed the day with retreat sounding and the lowering of the flag. Throughout the day, various bugle calls marked time and tradition, just as they had since the army's founding.

In the distance, many days, and some nights, we heard the dull thuds of tanks firing on the ranges. These sounds became so familiar that they blended into the background—no different from birdsong, barking dogs, or the occasional honking car. Only visitors seemed to notice. In fact, the absence of those sounds was more striking than their presence. Even residents of the surrounding civilian communities had grown used to what I liked to call "the sounds of freedom." Many neighbors understood, as we did, that our safety and liberty are secured by the Army's readiness—and its ability to deter those who might do us harm.

Nannette had grown accustomed to her role as a senior military spouse and was comfortable both representing military families to the civilian community and participating in family member activities on post. Senior spouses in any command are usually key figures in various spouse organizations, programs, activities, and projects. They also serve as advisors to different groups. The Red Cross, Spouse Clubs, Army Community Service volunteers, and the USO were only a few of the many activities supported by these spouses.

There were also formal and informal social groups, often centered around games or crafts like bridge, bunko, or quilting. Nannette had learned the Far East game of Mahjong and was both enthusiastic and skilled. She found other spouses who shared her interest and recruited novices to whom she taught the game.

I would sometimes come home for lunch or stop by in the afternoon to pick something up—easy to do, as my quarters and those of the CG were close to our respective headquarters buildings. I knew all the senior spouses, particularly those in the Mahjong group, and enjoyed feigning disapproval when I found them engaged in vigorous games in my quarters. Casting a suspicious eye about the room, I would declare "I see you ladies are playing that communist game again!" They looked forward to my mock disapproval and would even ask if I had "converted" when I failed to make my usual remark.

In the year that sped by, another significant event touched the Jordan family: I became eligible for consideration for promotion to major general. Brigadier generals could be considered for promotion after successfully serving one year in grade, and I was surprised to learn—when the zone of consideration was announced—that I had made it by a single day. Had I been one number lower on the brigadier list, and been promoted one month later, I would have needed to wait an additional year just to be considered for this promotion.

Of course, being considered was no guarantee of promotion. Still, as is often said in sports, you have to compete to be in position to win.

Some people believe that things tend to go from bad to worse. I had always thought that, with hard work, things tend to work out for the better.

As it turned out, I was selected for promotion and about to receive a second star. While some congratulated me on my good luck, I couldn't help but remember what several of my mentors had told me: Luck is what happens when preparation meets opportunity. People make their own good luck by working hard and being ready when opportunity arises.

I also knew that none of us make it on our own. We all stand on the shoulders of those who proceeded us and helped us along the way. More importantly, we stand in the palm of the Master's hand. I thanked the Lord for the good fortune and blessings my family had received.

I continued to serve as Assistant Commandant. Serving in a position while in promotable status was nothing new to me—I had done the same as a colonel while awaiting my brigadier general star. This time, however, I didn't have to wait long. The Armor Center commander was soon nominated for promotion to three-star lieutenant general and received orders for III Corps at Fort Hood.

What many people don't realize is that major general or rear admiral is the highest permanent rank by law across all military services. Three- and four-star flag officer ranks are special appointments, designated for specific, authorized positions—usually for only two or three years. These appointments require presidential nomination by name and Senate confirmation.

When officers reach the end of the term specified for their appointment, they must be nominated for and appointed to another position of equal rank. If not, they revert to their permanent rank—usually two-stars. Most choose to retire rather than return to a lower rank.

None of that concerned me at the time. Once I was promoted to major general, I would be tenured for five years—longer than I needed to worry about. It was far more exciting and enjoyable to live in the moment. I wondered where the Chief of Staff would send me.

The answer came in another call from the Chief, General Sullivan. In typical fashion, he reminded me that he had told me not to screw up as Assistant Commandant—then, after a pause, added, "and you didn't." He opined that I was probably trying to guess where I was headed. The answer, he said, was "no place."

The Chief explained that he had many positions to fill and plenty of two-stars available. But there was only one Armor Center, and he needed someone who could steer the effort to digitize the Army. The commandant and commander of the Armor Center also commanded the Mounted Battle Laboratory, the lead agent for advancing the mounted force in new technology, doctrine, and tactics. I was to replace my departing boss.

The change of command was preceded by a promotion ceremony, officiated by my former and current commander—and most importantly, my mentor—General Fred Franks. As always, except for the promotion in the Saudi desert, Nannette assisted in pinning the new insignia rank on my uniform.

The third full workday in command began with a tragic occurrence. I had already met many of the civilian employees and had made a concerted effort to thank them for their work and contributions to the Armor School's accomplishments—a gesture that was appreciated by the workforce. That rapport only deepened as we faced what happened that morning together.

A disgruntled civilian employee at the Training Support Warehouse, struggling with difficulties at work and at home, inexplicably entered the workplace and opened fire on his co-workers. He killed three and seriously wounded two others. He then drove forty miles to a Veterans Administration hospital, where he shot himself. He died two days later without regaining consciousness, never revealing his motives.

The tragedy stunned everyone. Yet the care and concern shown to the survivors by a grieving community earned my deep admiration. The incident redoubled my commitment to care for the entire community—military and civilian alike.

As I reflected on the event, I was reminded that nearly every place I had served had experienced some form of tragedy, either in my unit or one nearby. It wasn't that such events were more frequent in military communities—though the work was often dangerous—but rather the sheer number of people, especially young soldiers sometimes prone to immature actions, that led to accidents. These tragedies also stood out more starkly because members of the military community felt such a strong bond with one another.

But this tragedy was different—more senseless, more pointless. I reminded everyone that all of us—leaders and co-workers alike—must watch and come to one another's assistance, no matter the issue.

Notwithstanding the tragic beginning, the next two years proved to be some of the most memorable and rewarding of my career in terms of contributing to the Armor Force and the Army as a whole. The Armor Center lead the Army's efforts to integrate technology and information systems into vehicles, units, and operations—a transformative effort the Army termed "Digitization."

Small, rugged computer were installed in tanks and other vehicles and networked with various systems to provide instantaneous, accurate positioning and communication. These systems included electronic graphical maps and operational schematics, enabling leaders and soldiers to view both friendly and enemy positions. Fortunately, the latest version of the Abrams Tank—the M1A2—came equipped with built-in information systems. The trick was installing and networking such systems across the rest of the force to give commanders a complete operational picture.

There were tremendous hurdles to surmount, as well as equally significant procedural and training challenges. To some, the initiative seemed right out of science fiction—an impractical dream. Yet, much like the emerging high-tech industry, the Chief of Staff demanded that leaders think outside of the box. The Army leaned heavily on American Industry's innovative capabilities, developing systems that integrated data from electronic sensors, video feeds, and drones. This information provided precise targeting data for tanks, artillery, attack helicopters, and fast-moving Air Force aircraft.

Still, leaders in TRADOC and across the branch centers understood that even with this new technology, the most important factor remained the human leader. Ultimately, success depended on commanders who understood the mission, made tough decisions to accomplish it, and, most importantly, cared for their soldiers and safeguarded their lives. The technology's true value was in enabling better, faster, and more informed decisions—it could never replace leadership.

The staff and I worked closely with other branches, communications, and acquisition experts across the Army, and with industry partners to build the prototype digitized Armor Task Force. That unit trained at Fort Knox and later participated in a National Training Center (NTC) rotation to test its performance under the most realistic ground-combat conditions possible.

This field exercise followed a series of training and evaluations using both constructive and virtual simulations. New tactics and procedures were developed to fully leverage the capabilities. It reminded me of the time I first brought the newly fielded Abrams Tank to NTC. The digitized task force could now execute operations with a speed and accuracy that had once only been imagined.

It truly was the stuff of science fiction, yet it was real. As the Chief of Staff remarked at the end of the NTC rotation, "Together we have made this real. We have seen the future. Years from now, we will be able to say we were here when the Army stepped into the future."

The family continued to be blessed. Karl was doing well at Syracuse and had one year remaining. His post-graduation employment opportunities appeared bright, based on the two internships with national companies during the previous two summers.

Nannette and I had become grandparents with the birth of Noelle, LJ and Erika's daughter. Everyone enjoyed having a darling little girl in the family, and it was truly a new experience.

Nannette's parents had grown accustomed to having us nearby, and I continued to admire and enjoy my father-in-law. Occasionally joining us at casual functions at Fort Knox, the Master Blaster had become somewhat of a hit among the senior NCOs at the Armor Center—reinforced by entertaining, if slightly apocryphal, stories shared by both me and Top Owens.

The NCOs once asked me how my new father-in-law, then an airborne first sergeant, and I had gotten along when I was a lowly second lieutenant. I answered that we had been on a first name basis: I called him "First Sergeant," and he called me "Hey, you."

The more truthful version was told by Top Owens. He told a group of NCOs that when I was promoted to major—after only eight years as a captain—he was proud of me and said so. Then he added, "Someday you might make a half-ass Sergeant First Class!"

It was just another humorous way NCOs express admiration. The line was remembered by the Fort Knox NCOs, who later presented me with a pair of Sergeant First Class chevrons as a parting gift.

The work of the Armor Center was going well and brought accolades—praise I deflected—crediting the hard work of subordinates. Yet in the middle of all the activity, Nannette and I found the opportunity and means to devote more time to each other.

After what newspapers called the "storm of the century"—which dumped 19 inches of snow overnight and froze solid in minus 19-degree temperatures—the post was called upon to assist surrounding communities with emergency generators, vehicles, and mobile heaters. Once the weather cleared and operations returned to normal, we decided to escape for a week of leave in a warm location.

Commanders always encouraged their subordinates to take leave but often didn't take it themselves. Perhaps they thought of themselves as indispensable. The problem was that if the boss does not take leave, many subordinates will be hesitant to take leave, fearing not being seen as working hard or not being a team player. I had always thought the way to encourage others was by setting the example.

There were other reasons, too. If you never took leave, your subordinates never had the opportunity to "fleet up"—practice running things in your absence. In war, everyone must be prepared to move up and replace a fallen leader. And without rest, you burn out. As I had seen at the NTC, a tired commander is a dumb commander.

That one week away was enough for me to return recharged and ready to work even harder. From then on, and for the rest of my career, Nannette and I planned a week of vacation each quarter. That ensured I used most of the thirty days of annual leave every soldier is authorized—and showed everyone that it was not only acceptable but wise to take time off. No one is indispensable.

Nannette and I came to enjoy traveling for pleasure rather than just on Army business. It marked the beginning of a tradition of several short vacations a year—a practice we would follow even in retirement.

Four months short of the expected three-year tour at Fort Knox, the inevitable orders arrived: We were moving again—this time back to Northern Virginia and the Pentagon.

As we began the familiar process of preparing, packing, and relocating, we reflected on our nearly three years at Fort Knox. We had reconnected with old friends and made new ones. One of the colonels commanding a Training Brigades was a West Point classmate. Both of us had served for twenty-seven years, yet I was a two-star general while most of my classmates still on active duty were colonels, with a few having reached brigadier general. Only Bill Nash, who had been on the same brigadier list, was also a major general.

I had indeed been fortunate—and once again, I thanked God for His many blessings.

As we drove away from Fort Knox for the last time, memories flooded our minds. It had been a truly memorable experience—a special place, a special time, and a special group of people. We couldn't help but feel a touch of sadness, notwithstanding the adventures that undoubtedly lay ahead.

Both sons were now permanently out of the house, except for occasional visits. Bandit had died just before Christmas. After thirteen years, it felt strange not to have his warm breath on the back of the driver's neck or the nudge of a cold nose as he leaned forward from the back seat. That dog had loved riding in the car—and helping the driver navigate.

As I drove out of the gate, past the Bullion Depository, I reflected on the true treasure I had found at Fort Knox. That treasure was the people we had met and served with—the warm, welcoming communities surrounding the installation—and the deeply satisfying work of training soldiers and helping shape both the present and future Army.

That was the real gold I had discovered on this part of my long route of surprises.

CHAPTER 18

More Surprises

THIS THIRD ASSIGNMENT to the Pentagon felt different from the previous two. It was 1995—thirteen years since my first tour there and seven years since I had last left. As a general officer, I was now authorized senior quarters on one of three Army installations in the area, which would be miles closer to the Pentagon—a tremendous consideration given the region's notoriously bad traffic.

The home in Montclair was comfortable but with the children no longer living at home, it wasn't necessary. The additional thirty- to forty-five-minute rush hour drive ultimately tipped the decision: We kept the Montclair house rented and moved into quarters at Fort Belvoir. The colonial-style general officer quarters were both convenient and spacious enough to accommodate visiting family. Plus, it was not a bad deal to have renters paying my mortgage.

During her last two stays in Northern Virginia, Nannette had worked in dental offices. While I served on the Army Staff as a major, she worked for a civilian dentist in the area. Later, while I attended the National War College and served on the Joint Staff, she worked in the Pentagon Dental Clinic. When I was assigned back to the Pentagon after the War College, I thought carpooling with my wife would be both pleasant and economical. However, the Pentagon Dental Clinic had fixed hours and closed promptly at 1630 hours (4:30 p.m.). By 4:45 p.m., Nannette would be impatiently waiting for her ride home.

Unlike the long and crazy hours I had previously experienced in the Army Staff Operations, the Joint Staff generally kept more reasonable hours. Still, leaving at 5:00 or 5:30 p.m. felt early to me, but was clearly not early enough for our carpool to succeed. It didn't take long for Nannette to join a vanpool and for me to resort to solo commuting and whatever crazy hours the job required.

During this third time in The Building, Nannette chose not to work. Instead, she devoted her time to various volunteer efforts, stayed active in Toastmasters, and joined a Mahjong game when one was available. She was also perfecting her grandmother skills, as LJ and Erika added a brother to join the young lady who was their firstborn.

I had been told I would serve as Assistant to the Army Personnel Officer—the three-star general who held the top position in the Department of Human Resources, officially known as the Deputy Chief of Staff for Personnel, or DCSPER. That assignment came as both a surprise and a puzzle, as I had no formal experience or training in personnel. I would have much preferred a return to Operations, where my experience resided. But as the old Army saying goes, "Do the job you're given; not the one you want." So, I threw myself into the new environment, determined to learn quickly.

Then came the first of two big surprises. I ended up serving just seventeen workdays in the position. What I hadn't known was that I was basically in a holding pattern while being vetted for a very different job. The Inspector General of the Army—a three-star position—reports to the Secretary of the Army and is directly responsive to the Chief and the Vice Chief of Staff. Known as The Inspector General, or TIG, the role involves examining, inspecting, and reporting on the efficiency, discipline, and readiness of the Army to its most senior leadership.

The Office of The Inspector General also conducts detailed inspections in areas such as nuclear and chemical operations, highly classified programs (sometimes called "black projects" because they rarely saw the light of day), and special mission units. Two additional responsibilities fall under the IG's portfolio.

The first responsibility is assisting soldiers and leaders who were struggling with bureaucratic errors or misapplied processes—work that closely resembles the responsibilities of command and supports both individuals and institutional improvement. The second is conducting investigations into alleged misconduct by senior personnel—general officers and civilians in the Senior Executive Service (SES), the highest levels of Civil Service. Though often unpleasant, this work is necessary to ensure leader accountability and the force's confidence in its leadership.

I was surprised to learn that I was being considered for the position of Deputy to the Inspector General, or DTIG. Several agencies had to conduct background checks to ensure there was no adverse information that might disqualify me from such a sensitive role. In addition, the Chief of Staff, Vice Chief of Staff, and Secretary of the Army would all need to approve the selection.

Because the IG and DTIG are responsible for conducting senior-level investigations—some of which can determine whether the Senate confirms officers for advancement—the staff of the Senate Armed Forces Committee also wished to review the nomination. I made it through all gates and was appointed Deputy to the Inspector General of the Army.

While the Inspector General's Office had broad responsibilities, the primary role of the Deputy was to oversee the investigative process. Every officer takes the same oath to "support and defend the Constitution of the United States against all enemies, foreign and domestic," repeating it with each promotion. But Inspectors General take a second oath, swearing to "accept the special obligations and responsibilities of the position freely . . . and without prejudice or partiality, discharge the duties of the office." That oath, printed on a certificate, hung on the wall of my new office.

As DTIG, my job was to oversee and supervise sensitive investigations—often involving senior officers I knew personally, had served with, or who were even classmates. When particularly difficult or personally disturbing cases arose, I sometimes walked over to the wall and re-read that oath. Next to it hung a copy of the West Point Cadet Prayer, which reads in part, "Help me to always choose the harder right than the easier wrong, and never be content with a half-truth when the whole can be won." Both served as daily reminders and guiding principles in my decisions and actions.

There were approximately 350 active-duty generals, 200 more in the Army National Guard, and 150 in the Army Reserve. In addition, there were probably 200 Senior Executive Service (SES) civilians. This population—about 900 individuals in total—received a number of misconduct allegations each year.

One might ask why any allegations would be made against such a senior group. The reasons for allegations varied. With over a million soldiers across the active Army, Reserve, and National Guard, and thousands of civilian employees—all of whom had a general or SES somewhere in their chain of command—any time someone felt mistreated, genuinely or otherwise, a senior leader was often included in the complaint. In the eyes of many junior personnel, the general must have known what was occurring and was therefore complicit. Another reason was the widespread assumption that people in positions of power or authority inevitably exploited their roles for personal gain or pleasure, whether or not that was true.

However, the statistics showed a very different picture. Year after year, an average of 94 percent were proven unsubstantiated. Of the remaining 6 percent, half were attributable to inadvertent mistakes or minor issues that caused no harm to individuals or the institution. The final 3 percent involved serious, and often ugly, misconduct.

Teams of senior investigators gathered evidence and sworn statements to evaluate each case. Any criminal activity was turned over to the Criminal Instigative Command. Inspector General (IG) findings went to the Vice Chief of Staff, who determined corrective action or punishment in consultation with the General Counsel and the Judge Advocate. I didn't enjoy the role of DTIG, but I believed it was extremely important. It was never pleasant to notify colleagues that they were under investigation or to approve findings of wrongdoing. Still, the process helped safeguard the professionalism and integrity of the Army's senior leadership.

I took heart in knowing there were few senior leaders who actually behaved inappropriately—and that our efforts often cleared the names of those wrongly accused. I was proud of the vast majority of my compatriots.

That didn't stop my sense of humor from sneaking in. Like many in the Pentagon, I carried notes and papers in a portfolio. At the time, the television show *The X-Files* was popular, and I placed a mock cover on the portfolio that read, "*The X-Files*; The truth is out there." Curious generals would sheepishly glance at it—some even asked if they were in the folder. I sometimes replied, "No. But I can add you if you want."

The emotional weight of the job was often lightened by the Army trademark dark humor. It also helped to have my old lawyer friend, John Altenburg, serving as Deputy TJAG—Deputy The Judge Advocate General of the Army. Alongside my dear friend the Deputy General Counsel, we reviewed each case to ensure that due process was observed and that all regulations were properly applied.

We often joked that the only other senior folks who wanted to be seen with us were the Chief of Chaplains and the commander of the Criminal Investigative Command. There was probably a grain of truth in that.

For Nannette, the role of a senior spouse changed somewhat in Washington, DC. There were fewer activities tied to one's specific staff organization compared to operational units or TRADOC Centers. Most official functions were hosted by the Chief of Staff or other very senior generals, so participation was usually in the role of guest. However, the Chief of Staff received so many invitations that two- and three-star generals were often detailed to represent him. As a result, we frequently represented the U. S. Army at various events.

The District of Columbia offered a plethora of opportunities. Nearly any interest or activity could be pursued—whether theater, art, museums, or involvement in social and professional groups. One particularly memorable event was an embassy visit. Unlike the typical lavish evening affairs attended with spouses, a few other generals and I were invited to tour the new Russian Embassy during the day. The building had gained notoriety during construction due to the presence of unusual antennas—suspected of housing espionage equipment—and Russian charges that the U.S. had embedded listening devices in the structure. The truth was difficult to pin down, but the controversy was typical of the hyperbole exchanged between political adversaries.

Standing in the embassy's grand, overly opulent ballroom, we were struck by the contrast between its extravagance and the poor housing conditions known to exist for most Russian citizens. Though the communist regime had fallen, and Russia now spoke of "democracy," the legacy of old attitudes lingered. When we commented on the grandeur, our Russian escort officer smiled, winked, and said, "Yes. The last great gift from the grateful Proletariat."

At last, a Russian with a sense of humor. The moment reminded me of the Soviet guard I had once encountered at the East German border. You could strip a Russian of his communism, but you could never rid him of his Russian attitude or culture.

At the end of the first year back, Karl graduated from Syracuse, and, like his brother, entered the workforce—pursuing a career in corporate insurance. Another major family event occurred the following autumn while LJ and his family were visiting. It marked the second big surprise, one certainly as momentous as the first.

Five years and three jobs after graduating from college, LJ announced that he wanted to join the military. The news was completely unexpected, but welcomed. When I asked if he had made any firm decisions, LJ stated that he planned to visit Quantico Marine Corps Base that afternoon to see what they might offer. I calmly asked where he planned to have dinner that night. Puzzled, LJ replied that he planned to return and eat with the family. I barely looked up and replied, "Your family is certainly welcome to dine, but if you plan to join the Marines, you'll have to eat elsewhere."

Everyone laughed, but LJ got the message. Deep down, he knew the Army was the better route for him. A few months later, he enlisted and began Basic Training at Fort Jackson, South Carolina.

Since he already held a Bachelor of Arts degree, he qualified for an enlistment option to attend Officer Candidate School (OCS) after successfully completing Basic. Several others in his training company had the same OCS path, and together they bonded through a shared effort to endure everything the drill sergeants threw at them.

The twelve-week OCS course was much tougher than Basic Training, with stricter discipline, more intense physical training, and rigorous classroom instruction on top of field training and tactical exercises. LJ was twenty-seven—just months younger shy of the maximum age for officer candidates. Some jokingly called him an old man, but his added maturity paid off.

He performed well in OCS and was designated a Distinguished Military Graduate. We proudly watched his class graduate, and I had the privilege of administering his oath of commission. Major General Larry R. Jordan swore in Second Lieutenant Larry R. Jordan Jr. into the Regular Army and the Armor branch. Another generation of our family had begun its service to the nation.

LJ reported to Fort Knox for the Armor Officer Basic Course and received assignment orders to the 3rd Armored Cavalry Regiment at Fort Carson, Colorado. Over the next three years, he would serve as a tank platoon leader, scout platoon leader, and troop executive officer—a great start to his military career.

LJ's entry into the Army also served as an impetus for me to begin thinking of the Army as the family business. I had long known it was the perfect career for me. Still, with the number of educators in my family, teaching had always seemed the tradition among kin. Even LJ had taught for a while during the five years between graduating from Wake Forest and entering the Army.

I had enjoyed my three years teaching history at West Point, but I had always seen it as an interesting and enriching interlude between the more serious business of soldiering and serving the nation. At the time, I was still unaware of any ancestor who had served during the Civil War, but I did know that with LJ's service, four generations of our family had worn the country's uniform. I also realized how little I knew about my family history and heritage and I resolved to fill in those many gaps. That, however, would have to wait. My duties, already substantial, were about to increase.

Near the end of my second year as DTIG, the officer serving as TIG retired. The Secretary of the Army selected me to replace him, and for the first time, I went through the nomination process for appointment to lieutenant general.

In addition to the usual background checks for adverse information, I was required to submit a financial disclosure statement. While I had a few modest investments and decent credit, I was surprised anyone would think my finances, built on a soldier's salary, might be of interest. High government officials, members of Congress, and political appointees routinely submit such disclosures to avoid conflicts of interest. I was told this was the rationale behind the requirement.

Still, it came as a shock when DoD lawyers informed me that I would need to divest a few shares in a railroad and a company that manufactured paper products, including diapers. The bigger issue was that Nannette, as my spouse, would also need to divest a few shares she had purchased in a growing tech company—even though they were solely in her name. The explanation was that these companies had government contracts.

I thought it was bureaucracy run amok, but if the nomination were to proceed, I had no choice but to comply. I asked if we could place the holdings into a blind trust, as politicians often do, but was informed that option wasn't available to me. It reinforced my belief that soldiers are truly held to a higher standard, and it gave Nannette some extra ammunition in discussions about the sacrifices of military spouses.

Fortunately, three-star nominations usually don't require testifying or appearing before congressional committees. The waiting, first for committee review and then for the full Senate vote, was the most agonizing part. In the end, I was confirmed without any problem.

▼ ▼ ▼

One change that both Nannette and I enjoyed upon my move into the TIG position was relocating from Fort Belvoir to Fort McNair, located in the District of Columbia. The designated quarters for The Inspector General was a huge, three-story Georgian-style home along a tree-bordered parade field. The house backed onto the Washington Channel, which emptied into the Potomac River.

It was one of ten such homes reserved for some of the Army most senior generals stationed in the capital. A few others, including the Chief of Staff and the Chairman of the Joint Chiefs of Staff, lived at Fort Myer, closer to the Pentagon. Fort McNair, however, was also close to The Building—just a seven-minute drive on average.

The homes along our street offered panoramic views of the Washington skyline from their backyards. At the far end of the parade field stood the National War College, where I had once been a student. I often ran past these very quarters during morning PT, wondering what those "old timers" inside were doing. Now I was one of those old timers.

Living at Fort McNair was idyllic. It was a quiet, secure enclave at the southern tip of the District of Columbia. In the evenings, we would sometimes stroll to the point and watch planes arriving and departing Reagan National Airport, or observe dinner boats and pleasure craft across the waters. When fireworks were scheduled in the District, especially for Independence Day, neighboring families would often gather for backyard picnics with a perfect view of the display.

Years later, we still laughed about one incident from our time there. After attending a formal evening affair, we returned home—me in full dress formal uniform, Nannette in a cocktail dress—only to realize we had had forgotten the keys. The door we had exited earlier had locked automatically.

Fortunately, a first-floor window had been left open. It was decided to boost Nannette through it so she could unlock the door from inside. However, the window was quite a ways off the ground, and just as I was in the process of tactfully pushing my wife up to it, a military police patrol happened to drive by.

Seeing someone apparently attempting to surreptitiously enter a set of general officer quarters, they stopped to investigate. Once they realized what was happening, they quickly offered assistance. I smiled and said, "Thanks, but I've got everything in hand."

My elevation to TIG expanded the scope of my daily responsibilities. With a new DTIG installed to manage senior investigations, I turned my attention to the office's other functions. These included regulatory oversight of sensitive classified programs, special mission units, and matters related to nuclear and chemical operations. These were fascinating yet demanding areas, requiring constant attention and vigilance due to their complexity and ever-changing nature.

Regular visits and inspections of Army units around the world offered a broad, unmatched view of the organization. Through my travels and reports from subordinate Inspectors General, it became clear that the Army was operating exceedingly well. For such a large and complex institution, there was remarkable cohesion. Most soldiers were motivated, hardworking, and proud of their contributions. Leaders were equally dedicated: working diligently, making sound decisions, and correcting problems as the arose.

While occasional failures made headlines and stirred concern, I focused on ensuring senior leaders had a realistic picture of the Army's overall health, which, in truth, was good. Not perfect, but good. In an organization of over a million people, the occasional shortcoming was inevitable.

I sometimes joked with close friends and colleagues, such as the Judge Advocate General or the General Counsel, by saying, "The Army is a big organization. Someone loses their mind every day, regardless of age, rank, or gender." It was simply a reminder that we were all working with imperfect human beings, because that's all we are.

▼ ▼ ▼

A number of major events occurred that required special investigations and inspections. These included systemic training accidents, incidents of sexual misconduct, and suspected extremist views held by a few junior soldiers. The resulting inquiries helped identify underlying problems and ensured that Army values were being upheld.

All in all, I found the position deeply rewarding. I felt that I had truly contributed to helping the institution live up to its stated values, and to serving the nation with pride and honor.

I also realized that my time in uniform was likely drawing to a close. I had thirty-one years of service, and during my interviews with the Secretary of the Army and the Chief of Staff, it was made clear that the position of Inspector General was traditionally the final assignment before retirement. That policy existed to ensure the incumbent could prioritize objectivity, pursuing duties without concern for how decisions or actions might affect future career opportunities.

I had chuckled to myself at the thought. As a lieutenant, I had once been told by a captain, who had received a battlefield commission and didn't think much of "college-boy officers," that I would never make captain. Being appointed a lieutenant general and Inspector General of the Army, even if it was to be my last job, was more than enough satisfaction.

By that point, I had served in the role for two years and anticipated retiring after one more. Nannette and I had already taken a week of leave to visit a guest cottage at an Air Force base on the Florida coast, where we began thinking about retirement—what we might do, where we might live, and how we'd spend our new freedom and time. We came away with a rough plan.

While on an official visit to my old stomping grounds at Fort Knox, I received a call from the Chief of Staff. By then, the Chief was General Ric Shinseki, who had been a First Classman in the West Point Class of 1965 when I was a Plebe. A fellow armor and cavalry officer, we had been friends for years, and I had followed him in command of the brigade in

Kitzingen, Germany. We often spoke fondly of the brigade, the Kitzingen community, and regularly greeted each other with the brigade's motto "Send Me!"

I was surprised he had tracked me down while I was traveling, and even more surprised when he said he needed a favor. He wanted me to accept another assignment, as an exception to policy, and return to Europe for a third time.

At the time, the commander of U.S. Army Europe (USAREUR) was deployed full-time to Sarajevo overseeing operations in the Balkans. His deputy, who had been managing the rest of the European theater from Heidelberg, had recently retired. The Chief wanted me to take that position and, in his words, "put your arms around the theater." The idea of continuing to serve, and returning to Europe, was joyfully received. I replied without hesitation, "Send Me!"

The usual nomination and confirmation process was underway, shepherded through the Senate by staffers I had worked with on numerous cases and issues. I had earned their trust and admiration through the agency's thorough and candid investigative reports on senior officers under consideration.

The saying went, "If the army IG says something, you can take it to the bank." That wasn't my doing—it was the result of a longstanding Army reputation for integrity in that area. My responsibility had been just to maintain it.

As I prepared to depart the Inspector General's Office, I did so with a deep sense of pride and accomplishment. The last items I removed from my office wall were the Cadet Prayer—featuring the line, "Help me to always do the harder right rather than the easier wrong"—and the certificate bearing the Inspector General oath.

CHAPTER 19

A Final Tour

SITTING IN THE airliner cabin over the Atlantic Ocean, we reflected on how different this trip was from the first, twenty years earlier. We did not have two restless boys to watch. There were no cargo planes with "comfort pallets." And there were no more mysteries about what lay ahead. By now, we were what people called "Old Germany hands." We had lived there a total of seven years and knew we would enjoy the experience. The food, the wine, the people, the culture, the travel—it was all familiar and had been deeply missed, even though we had enjoyed our time back home. We were as comfortable in Germany as in Virginia or Texas. Life was still an adventurous journey, and once again we were on that road.

One of my West Point classmates, now a two-star, picked us up at the Frankfort airport in a hard car—a specially equipped, armored Mercedes sedan with bulletproof glass, run-flat tires, and other security features. Ever since General Kroesen's car was attacked in 1981, when then-Major Jordan was a War Plans Officer in Heidelberg, senior generals had been provided hard cars. The drive to Heidelberg was swift as we sped south at around 140 kilometers per hour, or roughly eighty-six miles per hour. With no speed limit on most of the autobahn, other cars occasionally flashed by even faster.

This particular hard car was an older model, but the office was soon to receive one of the newly-issued replacements—a BMW 750 sedan. It was more tank than car, weighing in at 6,000 pounds (three tons) thanks to its full armor package. Powered by a supercharged V12 engine and cooled by dual radiators, it could cruise easily above 190 kilometers per hour, or nearly 120 miles per hour. Drivers of these vehicles attended special defensive driving course and relished the thrill of flying down the autobahn. Just how fast the hard car could go in an emergency, I preferred not to find out.

Over time, Nannette was captivated by the chase car, a second vehicle carrying the Personal Security Detail, managing to stay within three feet of the hard car's rear bumper, no matter the speed, weather, or traffic.

▼ ▼ ▼

The quarters we occupied were the largest in the housing area—coincidentally, the same neighborhood where we had lived during our previous time in Heidelberg. It was one of the homes I had run past as a major, never imagining I would one day live there. In fact, this marked the third set of general officer quarters I had once seen and never dreamed I would inhabit—but ultimately did. The houses at Fort Knox, Fort McNair, and now Heidelberg held that distinction.

The pace and activities at USAREUR Headquarters were fast and varied. In addition to routine training, host nation engagement, and community affairs, the command was also responsible for supporting the substantial U.S. force deployed in the Balkans.

The USAREUR commander, General Monty Meigs—a friend, fellow National War College classmate, and West Point graduate one year ahead of me—was deployed full-time in Bosnia. I was tasked with spending about one week each month in Bosnia, Kosovo, or both.

During my two-year tour in Germany, I flew back to the United States a total of thirty times for meetings or official functions. Crossing the Atlantic became a routine. In addition, travel throughout Europe—especially Italy, Belgium, and Britain—added to the operational tempo.

Whenever possible, I brought Nannette along. Sometimes she was specifically invited by our hosts as a representative of the American forces in Europe, accompanying me in an official capacity. She especially enjoyed trips to Italy, where business took us to Rome, Vicenza, Naples, and Venice. All but Venice hosted major headquarters for NATO, the U.S., and Italian forces.

One of the things we always found particularly interesting when traveling together—especially to Italy—was the level of security provided due to my position. It was something only shared with family and close friends, because mentioning it to casual acquaintances might come off as egotistical bragging.

Whenever we arrived in Italy, we were met at the airfield and escorted to our destination by several motorcycle-riding police officers, the *Carabinieri*, or National Police. These officers rode large BMW motorcycles with impressive skill and flair. Often, they would cruise into intersections, rise to their feet while steering by shifting their weight, and signal traffic to stop with both hands; they seemed to delight in their daredevil maneuvers. We marveled at how they expertly ushered our motorcade through the chaotic streets of Rome with an ease and confidence that never failed to impress.

Another fond memory involved the Personal Security Detachment (PSD) that accompanied me everywhere while traveling on official business. My six-person team included a trained driver, a chase car driver, and four protective specialists. Dressed in civilian business attire and armed, they were highly professional and exceptionally skilled. I disliked referring to them as bodyguards, but in essence, that's what they were. Security protocol dated back to the attack on the USAREUR commander during my first tour in Heidelberg and reflected the heightened global threat of terrorism.

Over time, a lighthearted tradition developed. Whenever our schedule allowed for a break, Nannette and I would try to slip away for a few hours—dressed casually, blending in with tourists—to enjoy a quiet dinner for two. The challenge was whether we could avoid detection while the PSD team continued their protective surveillance without being seen. Without fail, the next morning the PSD team chief would casually mention, "Hope you enjoyed your dinner." This made it clear that our escape had once again been less than successful.

We also visited France to participate in ceremonies at American military cemeteries and attend other commemorative events. Trips to the United Kingdom included both official business and personal travel. One highlight was riding the train through the Channel Tunnel, or "Chunnel," as many Europeans called it, traveling under the English Channel from France to Great Britain.

Spending time in London reminded us of a trip we had taken from Fort Knox in 1994 to visit the British and German Armor Schools. We happened to be in Britain on June 6, the 50th anniversary of the D-Day Invasion. That day, I served as the senior American officer in the United Kingdom and represented the U.S. at a commemorative ceremony at one of the embarkation sites. While I was doing that, Nannette was invited to christen a special train named in honor of the port where many American troops had departed for Normandy. That moment, and many others like it, became cherished memories.

Still, we rarely spoke of such experiences outside our close circle of friends—those who would understand we weren't boasting, merely reflecting. Few outside the military world would have truly understood.

One of the major tasks I managed involved the movement and consolidation of American units in Germany. From a Cold War high of over 200,000 troops, USAREUR was drawing down to approximately 65,000—still a considerable force. However, not all the existing facilities were needed.

As the consolidation and restationing plan was finalized, more than 220,000 soldiers had to be relocated. Given the number of married soldiers, an estimated 25,000 family members were also affected. This required far more than just housing—schools, chapels, medical facilities, post exchanges, commissaries, and a wide range of support services had to be arranged at new locations.

It was a monumental effort. And throughout the process, training had to continue to maintain operational readiness. In many ways, it was like assembling a massive three-dimensional jigsaw puzzle. Fortunately, there was time to enjoy the European culture we had so missed. We delighted in attending the many fairs, fests, volksmarches, and other local events. We especially looked forward to the wine festivals, the annual Octoberfest, and the festive Christmas season with its traditional outdoor markets, decorations, town center band concerts, and hot spiced cider.

The Germans, in particular, seemed to embrace every opportunity to celebrate—whether marking a change of seasons, a holiday, or simply coming together as a community. Their American guests often adopted these traditions as their own.

Adding to our family memories, Nannette's parents were once again able to visit and enjoy Europe. Heidelberg was a new experience for them, as they thoroughly enjoyed taking in the sights and scenery of the historic town.

As has been said, this tour in Germany, more than any before, involved extensive travel. One trip in particular stood out.

U.S. European Command (EUCOM), the joint headquarters for all U.S. forces in Europe—and at the time, also responsible for Africa—had committed to a visit to Ghana as part of the American engagement strategy for the region. When neither the commander, an Army general who also served as Supreme Allied Commander for NATO, nor his deputy, a Marine general, were available for travel, a request was made for U.S. Army Europe to represent the United States. I gladly accepted, having never before visited the African continent.

The nearly twelve-hour journey aboard an Army corporate jet was so long that it required two aircraft and two crews, with a switch in the Azores for refueling and crew rest. The length of the flight brought home the vastness of Africa, especially considering I had only flown halfway down the continent.

The arrival ceremony was typical for an official visit to the chief of an allied army. An honor guard, military band, and welcoming delegation awaited at the airfield. Ghana's former status as a British colony was evident in the uniforms, music, ceremonial formality, and the anglicized English accents of the welcoming officers.

As salutes were exchanged during the playing national anthems, my Ghanaian host and I casually appraised one other. Shortly after meeting and beginning the visit, we each formed a favorable opinion of the other—what would become the foundation of both a professional and personal friendship.

Educated in the U.K. and gifted with a gregarious personality, the Chief of the Ghanaian Army held the rank of lieutenant general, like myself, but occupied a higher relative position as the senior army officer in his country. His easy manner and outgoing nature were complemented by a sharp intellect and solid military competence. I was impressed.

The visit offered the opportunity to experience both the military and cultural aspects of Ghana. I emphasized the importance the U.S. placed on Ghana as a model of efficiency, democracy, and political stability for the rest of the continent. We discussed areas where our two armies could expand training opportunities and deepen cooperation.

Equally impressive was what I observed among the civilian population. Ghanaians struck me as friendly, industrious, and ambitious—sincerely focused on education, national progress, and family. Signs and posters promoting the value of education were common in both urban and rural areas. Schoolchildren in British-inspired uniforms could be seen walking to and from classes. Businessmen and businesswomen moved purposefully through the streets. Elders were treated with visible respect. Ghana embraced modernity and technology while holding firmly to its traditions and cultural customs. It was a country looking ahead—without forgetting where it had come from.

A few days into the visit, I received an invitation to an informal dinner with the Ghanaian Army Chief, his wife, and a few close friends. The U.S. attaché stationed in Ghana was visibly excited, explaining that in his two years there, the Army Chief had never invited a visiting foreign officer to his personal home. He considered it a diplomatic and international breakthrough. I, on the other hand, saw it simply as a gracious courtesy extended by one senior officer to another.

I found it to be one of the most memorable of evenings. What struck me most was how similar it felt to so many experiences I'd had with close friends at home. The wife of the Ghanaian Army Chief was a bright, cheerful, well-educated woman—independent-minded, confident, congenial, and supportive of her husband. At the same time, she was not overly impressed by their social status, viewing it as a fortunate responsibility. She reminded me of Nannette and her down-to-earth approach to being a senior Army spouse. That impression was cemented when the hostess recounted the story of her husband's promotion to brigadier. He had been hoping for the news for months, and one day she saw him gleefully strutting home. Assuming he had received the long-awaited news, she met him at the door and said, "I saw you strutting home and assume you're going to be a general. I'm happy for you, but you can just leave that attitude at the door!" She was clearly the chief inside the house. I told my host that such things transcended national boundaries, and that in my own home, my spouse was the commander-in-chief.

The dinner gathering included three other couples—longtime friends of my host. A brigadier, two colonels, and their wives rounded out the group. All had known each other for years, and despite their differences in rank and responsibility, the atmosphere was relaxed, with friendly banter and teasing freely exchanged.

I remarked on how similar it felt to time spent with close friends back home—how familiar and comfortable the setting was. My Ghanaian hosts were delighted to hear this and shared that they had wondered whether Black American officers enjoyed the same kind of close personal relationships despite career distinctions. I assured them that we did. That evening, we spoke at length about the professional and cultural kinship that connected us—across ranks, roles, and national borders.

That sense of cultural kinship was powerfully reinforced during a visit to one of some Ghana's national historic sites at the end of the trip. I was taken to the coast to tour one of the old slave forts, where thousands of Africans had once been held before transport to the Americas. On the upper floors were the luxurious quarters of the slave traders. Below, in stark contrast, were the crude cells where the enslaved were confined.

A large wooden door at the base of the fort opened directly onto the beach, where captives were once rowed out to the waiting ships. Above the door was a carved sign that read Gate of No Return. It was a poignant and striking sight.

The Ghanaian Army Chief then asked me to step through the door. Once we were standing on the beach, he quietly asked if I could imagine the feelings of what might have been our shared ancestors as they crossed that threshold. Emotions filled both of us. We were two old soldiers, momentarily silent in shared reflection.

Then he asked me to turn around. On the outside of the door was another sign. This sign read Gate of Return. Some years earlier, archaeologists in South Carolina had recovered slave remains whose DNA was traced back to Ghana. Those remains were ceremonially returned through the gate and buried at the fort. The new sign had been installed to honor that moment.

As I stepped back through the gate, it felt as though I was returning, symbolically, to my ancestral homeland. I was a proud and patriotic American. And equally, I was a proud African American. To be reminded in Ghana of how far my family and ancestors had come, often in the face of unimaginable adversity, was deeply moving, and profoundly gratifying.

Two years in Heidelberg flew by, and we weren't eager to leave. But our departure was prompted by another call from the Chief of Staff—this time with another favor to ask. He wanted me to return to the Pentagon to serve as the Director of the Army Staff, known as the DAS.

The Secretary of the Army, Tom White, was preparing to reorganize the Army's leadership structure. The Army Staff, which reported to the Chief of Staff, and the Secretariat Staff, which worked for the Secretary, were to be brought into closer alignment, perhaps even partially consolidated. The DAS would play a central role in that effort.

Interestingly, Secretary White was the same Tom White who, twenty-three years earlier, had called me at Fort Leavenworth to inform me of my upcoming assignment in Germany. At the time, people said that Tom White knew everything important going on in the Army. I had doubted it then, but not anymore. Knowing the Chief of Staff well, I replied "I don't lie awake at night dreaming of being the DAS—but . . . Send Me." I was ready to serve if that's what the Chief of Staff needed.

The normal nomination process proceeded as expected. Once the nomination left the White House, it was sent to the Senate. Following the standard committee reviews, Senate rules required that the nomination "sit on the Senate floor" for at least twenty-four hours before a vote could be taken. While awaiting that vote, I was surprised to learn that the Army had suddenly withdrawn the nomination. I was puzzled. Had something gone wrong? Was I about to retire sooner than expected?

My concerns were quickly put to rest by another call from the Chief of Staff. He said he needed a really, really big favor. Unexpected personnel changes had created a vacancy for the Deputy Commander of TRADOC, at a time when the Army was undergoing its most significant reorganization since the mid-1950s. New systems and formations were being designed and integrated, and my experience at Fort Knox with digitization made me a strong fit for the role. This new assignment took priority over the Army Staff reorganization at The Building. I was genuinely more excited about the opportunity at TRADOC but contained by enthusiasm, responding simply with "Send Me!"

Some people say that bad things happen in threes. I prefer to think that good things often come in threes.

My third nomination to serve as a three-star general breezed through the Senate confirmation process. I suspected the staffers I had worked with during my time as Inspector General helped move it along smoothly.

I had just completed my third, and thoroughly enjoyable, tour in Germany, following my third and most satisfying tour in the Pentagon. I was now heading into my third assignment in TRADOC, having been there when it was first established. I was about to serve in my third position as a lieutenant general. And, as if to complete the pattern, I now had three grandchildren—LJ and Erika had welcomed a second daughter. Good things really did come in threes.

Our move to Fort Monroe, where TRADOC was headquartered, coincided with globe-rattling events. The threat posture in Europe had already been elevated for months, as intelligence pointed to possible terrorist activity across the continent. Several incidents, including a suspected kidnapping and multiple bomb threats, had USAREUR on high alert.

The day our household goods were packed and picked up seemed like any other. That afternoon, the moving van pulled away from our now-empty quarters. Nannette went to borrow a vacuum cleaner from a friend. With the radio and television already packed, we had been unaware of what was unfolding. Nannette returned with the devasting news: planes had struck the World Trade Center towers in New York and the Pentagon. It was September 11, 2001.

The next few hours were seared into our memories, as they were for all Americans. Our first thought was to call our son Karl, who lived and worked in Manhattan. Before leaving for Germany, we had visited him, seen his apartment, and toured his office. His office was on the twenty-third floor of a building in lower Manhattan, just a block from the World Trade Center. His window faced the towers.

The phones were jammed. From Germany, it was impossible to reach any numbers in New York City. Hours passed in anxious silence until we received an email from Erika, who was at Fort Lewis, Washington with LJ, who was a captain by then. Karl had finally managed to reach his brother by phone to let him know he was safe. Fortunately, he had a meeting that morning in Midtown, several dozen blocks north of Ground Zero.

We learned that our youngest son, nearly 4,000 miles away, was safe. We learned he was safe thanks to a message from our daughter-in-law, who was about 7,000 miles away, on the West Coast. As we marveled at the modern communication, I received another call—this one came from Johnston Island in the Pacific, roughly 750 miles southwest of Hawaii.

It was the younger daughter of our friends Larry and Jean Ellis. She worked for the Army and Air Force Exchange System (the PX) and had just heard the news. She couldn't reach her parents and was deeply concerned. Her father, a lieutenant general, served as the Army Operations Officer (DCSOPS) and worked in the Pentagon. The plane had struck the Army side of The Building.

Worried and unable to reach them, she called her father's longtime friend, me, on the other side of the globe. I was able to reassure her that both her parents were safe. I knew that they were vacationing in Jamaica at the time. As always, the circle of friends had kept each other informed and connected.

We arrived in the States a week later, landing at Washington Dulles International Airport. On our way to the hotel, we passed the damaged Pentagon. A flash of anger surged through as I took in the sight of the destruction—an attack on a symbol of American strength. The world had changed. So had the lives of all Americans.

Every soldier felt a sense of anger that an enemy had struck us in such a bold and devastating way. 9/11 became one of those moments indelibly etched in memory, where everyone remembers exactly where they were when it happened. For me, it was like the day President Kennedy was assassinated. I would always remember where I was when I heard the news.

CHAPTER 20

The Last Bugle Call

COMING HOME TO the States after previous tours in Europe was always a joyous occasion. This time, however, was different for two reasons. First, there was a strong sense that we wouldn't see Europe again for a long time. Second, the nation was blanketed with anxiety and bitterness following the 9/11 attacks. People seemed uncertain, fearful, and mentally and emotionally displaced.

When we reported to Fort Monroe, Virginia, we found the installation locked down, its gates fortified as if General Robert E. Lee's forces were about to attack—like the dark days for the Union in 1862 during the Civil War. After witnessing a very real ground threat in Europe over the past two years, I thought some of the heightened security measures were overkill; or, I thought perhaps the measures the Fort was taking were simply designed to give people a sense of security.

Arriving at Fort Monroe was also unlike any of my many other arrivals at new posts over the past thirty-three years. As old Cavalry troopers would have said, this was to be the last bugle call I would answer. After this assignment, it would truly be time to retire. I had one last job to do, and I planned to do it well.

Still, I couldn't help but look ahead with some trepidation to the day when the uniform would come off for the last time. Although it didn't dominate my thoughts or time, Nannette and I often talked about it, along with the three decades of memories that would always stay with us. Looking back, we could hardly believe all we had done, seen, and experienced. At the same time, we were struck by how quickly those more than thirty years passed. It had been quite a journey together, an odyssey, on a route of surprises.

Thinking back to the transition planning we had done years earlier when I was The Inspector General, believing it would be my final assignment, we dusted off those old plans and found them still valid. At the same time, we remained committed to continuing the journey to the end, with the sense of purpose, humility, dignity, and adventure that had sustained us until then. That was part of being an Army family.

Awaiting us when we arrived at Fort Monroe was our enlisted aide, Sergeant First Class Robert Canada. He had already been with us for four years, both at Fort McNair and in Heidelberg. Sergeant First Class Robert Canada—or "Sergeant C," as Nannette called him—was responsible for keeping my uniforms in good shape and ready to wear. He also maintained the public areas of our quarters used for official entertaining and prepared meals for those occasions. He had flown back from Germany with his family early to prepare for our arrival.

Sergeant Canada and his family had truly become part of the Jordan extended "Army family," having proven themselves as a resilient and professional Army family. Sergeant C appreciated the unpretentious way we treated him—not as a subordinate, but as a team member. When he initially interviewed for the job, I found him qualified and satisfactory. However, I told him to meet with Nannette first, since he would be around her more during the day and they would need to get along. I joked that my wife, the daughter of an Airborne First Sergeant, was "Old Army." She was very much in charge of the house and I was only an occupant.

Later, Sergeant Canada admitted he had been both surprised and a bit apprehensive about the casual and candid way the first three-star general he personally encountered had come across. This was not what he had expected. I explained that Nannette and I were open and approachable people, but we expected our private lives in the quarters to remain private. Some aides had a reputation for gossip, but Sergeant C understood the trust we place in him.

The mutual respect helped form a relationship that was probably not common between a general, his spouse, and the enlisted aide. It was professional relationship, yet also friendly and personal. I had similar relationships with my officer aides-de-camp, who were often not much older than my own son in the Army. However, I rarely kept an officer aide for more than a year, as I believed any longer was not good for their careers. In a year, they would learn much about how senior officers made decisions and handled situations. But I didn't want them to become "professional horse holders," making a career of ingratiating themselves to senior officers who could open doors for them. I didn't want that type of officer around anyway.

The good ones needed to grow from the experience and return to troop duty to practice their branch and leadership skills. Still, I maintained long-term relationships with most of my former aides.

The Army was in the process of organizing itself for the next twenty years by fielding new equipment and developing new formations and doctrine. It reminded me of what had happened during my time at Fort Knox. Keeping the TRADOC staff synchronized and working productively took up much of the deputy commander's time and energy.

Another task that occupied me occupied was synchronizing the support needed to train soldiers at the twenty-nine different schools.

These schools ranged from Basic Training for new recruits to flight school for aviators to the Command and General Staff College for field grade officers. The Army sometimes changed the number of soldiers to be trained in each specialty during the year. However, the Department of the Army did not always seem to realize, or truly understand, the amount of work required to transfer funds, personnel, contracts, or other support between dispersed installations to make those changes within a given year.

▼ ▼ ▼

The two years living at Fort Monroe were extremely enjoyable. Our quarters were on a street with senior officer homes facing a parade field with a band gazebo, beyond which lay the Chesapeake Bay. Across the bay was the Naval Station Norfolk base, and watching the fleet depart and arrive became a favorite pastime. By the end of our tour, we could identify nearly every type of major naval vessel.

In the warmer months, the band gazebo hosted band concerts that we could enjoy from our large front porch. On weekends, it was often the site of weddings, which we sometimes discreetly observed. We often remarked that the stately old three-story set of quarters would make a charming bed-and-breakfast. Like many senior couples whose children were grown and gone, we joked that we could have used the space when our kids were younger, but now we only used it for official entertaining.

Time passed quickly as the work of designing the future Army continued. My good friend Larry Ellis and I were on the same continent for the first time in nearly a decade. Ellis, who had served as DCSOPS in the Pentagon, would soon become the Forces Command Commanding General and earn his fourth star. We communicated regularly and saw each other whenever possible. Most of my close friends had already retired, though we still maintained contact.

Sergeant C had been promoted to Master Sergeant, for which Top Owens heartily congratulated him. LJ received a promotion, completed the Captains Advance Course, and was assigned to a unit at Fort Lewis, Washington, equipped with the new Stryker vehicles—wheeled armored fighting vehicles. Nannette was fairly knowledgeable about the various vehicles in the Army but took special interest in the Stryker.

While walking through a display of equipment at the annual exposition conducted in Washington, DC, by the Association of the U.S. Army, she asked the manufacturer's representative on the floor for a tour and explanation of the Stryker vehicle. After several minutes, she dismounted the vehicle and turned to me, saying, "I guess that will be okay for my baby to go to war in." I was amused and replied, "When that boy's daddy went to war, you didn't seem as concerned." She quickly put me in my place, responding, "I could replace his Daddy." Army wives shared a unique sense of humor that was part of the Army community.

LJ would soon take command of a troop, or company-sized unit, and deploy to Iraq. Before his deployment, he occasionally called me at work to ask questions, seek advice, or share updates about his growing family, which now included three daughters and one son. One day, my secretary announced that Captain Jordan was on the phone and "wanted to speak if the general had time." This had become a fairly regular occurrence. Although I was busy and not prepared for a long chat, I answered the phone, planning to suggest calling me later that evening if it could wait. After quick pleasantries, LJ complained about a policy recently announced by the Department of the Army and asked, "What are you guys doing?" By which he meant, what are you generals up to? Are you Old Goats crazy?

I calmly asked, "Are you standing up?" He replied that he was seated at his desk. I then said, "Stand up, and put your heels together." In other words, stand at attention. That conversation was over. I didn't have time to listen to a captain, even if he was my son, grouse about a policy he probably didn't understand the background of. A captain's job was to execute the policy.

Mostly, though, I just wanted to pull LJ's chain a little. I actually greatly enjoyed serving at the same time as my son and often told close friends that those six years we wore the uniform together were the best of my career. It felt like the family business.

In 2003 as the thirty-fifth anniversary of my graduation and commissioning as an Army officer approached, Nannette and I prepared for what would be our last move with the Army. I had long decided to retire while I was still having fun and enjoying the institution. I didn't want to wait until I became bored, dissatisfied, or simply too tired to do my best. Most generals retire before reaching thirty-five years of service. Nannette and I made the decision together. We had flown back from Germany in 2000 and sold our Montclair home after nearly nineteen years. We knew we would never live there again, and it was too far outside the District of Columbia to take full advantage of the city's venues in retirement. We had decided to remain in Northern Virginia, which had truly become home after four tours. Mentors advised us to move someplace we liked, rather than chase a job. Employment would come if I wanted it. We were also advised not to chase after our kids, as they would eventually move on. LJ certainly would, being in the Army, and Karl was already in his third location of his business career.

We were also advised to live somewhere our friends lived or would pass through. Washington, DC, and northern Virginia fit all those categories well. So, we purchased a fairly new home in Fairfax Station, about twenty minutes south of the District. Our new home was close to military facilities for use in retirement and convenient to the area's airports.

As I thought about the past thirty-five years, I reflected on the many places I had been, the people I had met, and the times I had experienced. Each was special. I remembered the highs and lows, realizing I had far too many blessings, wonderful memories, and high points to ever regret the few low ones.

I was proud that Nannette and I had remained humble throughout our successful journey, never considering ourselves VIPs or special. We both held on to a message shared during the Brigadier General Orientation, the so-called Charm Course. A wise spouse of a senior retired mentor had said being a flag officer and spouse was like children playing dress-up. When the game was over, you had to change back into your own clothes. The message had been clear: Don't change; stay true to who you are. I was thankful that we had.

Our last night at Fort Monroe, I thought of many of those things. As I lay in bed for the last time on an Army Post while on active duty, I heard the haunting notes of "Taps" being played, as it was every night. Symbolically, it was the last bugle call.

As expected, the retirement ceremony was bittersweet. It wasn't easy stepping away from the only thing I had done my entire adult life, but I was proud of my service and felt fulfilled. My friend, the Chief of Staff, General Ric Shinseki, officiated the ceremony at Fort Myer. The weather had been unpredictable, so the ceremony was held indoors in a large riding hall used for such events. The Army Band and Ceremonial Unit participated.

Many of our friends from across the years were there, as were both of our sons and our grandchildren. Much of the extended family and our closest friends made the trip to see an old soldier retire. Both Nannette and I received awards for our service from the Chief of Staff, and I was presented with a certificate of retirement and a flag that had flown over the Capitol.

In my farewell remarks I expressed appreciation for the opportunity to serve the nation and to lead its sons and daughters for three and a half decades. I shared that I had recently been asked what legacy I wished to leave in the Army. I told the attendees that the only legacy we can truly leave is in the people and places we touch for better or worse, and that I hoped to have touched them all for the better.

I thanked all the great NCOs I had worked with and learned from, including First Sergeant Owens, the Master Blaster, and my mother-in-law, Lillie, for being great role models. I thanked my sons for the sacrifices of growing up always on the move and never having one place to put down roots. Most of all, I thanked Nannette, saying it had been a great journey—but one that would not have been worth it without her at my side.

That evening, after the ceremony was long concluded, I thought about all that had occurred, all that I had seen, and all that I had been blessed to experience. I thought about how my family had shaped me, and about my son, who would continue what I called the family business. There was no way for me to know that in five years, my nephew John would join the family business as well, or that in fifteen, my grandson Brandon would, too.

That night, though, I thought about the service of my father-in-law, First Sergeant Dallas Owens. I remembered my father, Tech Sergeant Napoleon Jordan, and uncle, Tech Sergeant Herbert Johnson, who both served during World War II. Finally, I thought about my great-uncle, Private Joseph McKim, the Buffalo Soldier. The only thing that might have been more poignant would have been knowing, as I would learn seventeen years later, that my great-grandfather had served in the Civil War.

By that time, I would have completed a second career as a senior executive in a 500-person consulting firm, working alongside many old friends and associates from the Army, and had gone on to focus on nonprofit work with several service and volunteer organizations. After years of joking that I should write a book about my experiences, I finally began the task. Additional genealogy research then revealed that my great-grandfather had, in fact, served during the Civil War. That discovery helped refine the direction of the project. I decided to include the significance and impact of the six generations of my family who served as soldiers—something I considered somewhat unique for an African American family.

My long road march was coming to an end, but my family would continue it. However, that was in the future.

That night after my retirement, I also thought about the sacrifices of family and the role they played in helping their soldiers serve. As Nannette's friend Betty had once said of Army wives, "They also serve, who only stand and wait."

Epilogue

THIS WAS MORE than the story of my own military career. More importantly, it describes the route I traveled to discover the untold story of my family's military heritage and to see the Army as a true family business. It is a story that spans nearly a century and a half, beginning during the Civil War.

In retirement, I continued serving in the institutions I loved as chairman of the board of directors for the West Point Association of Graduates and as a director on the board of the Association of the U.S. Army. I also served on the board of the Army Historical Foundation, which built the National Museum of the United States Army. My non-profit work included organizations focused on mentoring young officers and promoting the development of minorities and women in the public sector. In addition, I spent ten years in the defense industry before finally retiring to devote more time and energy to volunteer work.

Nannette continued her volunteer work with the chapel and elsewhere. She eventually became a county election officer, taking pride in helping with the democratic process. Toastmasters, Mahjong, and other pursuits also helped fill her spare time, but she most enjoyed travel with me and spending time with our sons and grandchildren. As she would later write to me, "While not always a smooth road, we are still traveling it, and overall, it has been a good trip."

The story continued as Larry R. Jordan Jr. traveled his own path. Shortly after my retirement, LJ was deployed to Iraq. That first combat tour lasted fourteen months and marked the first time his family had faced such a long separation. It wouldn't be his last deployment. A second year-long deployment to Iraq was followed a few years later by a year in Afghanistan. Shortly after returning from that third combat tour, LJ led an advanced party of soldiers to Haiti for a humanitarian mission in response to a devastating earthquake. LJ later remarked that the misery and suffering he and the soldiers witnessed among the Haitian people had affected most of them more profoundly than the trauma of combat, a reminder that not every painful memory or lasting shock comes from combat.

Fifteen years after I retired—and twenty-one years into his own career—LJ's son, Brandon, entered the Army, becoming the sixth generation of our family to do so. Brandon would serve in Afghanistan as a combat medic, bringing home lasting memories, a sense of reward, and a humble satisfaction in being able to care for his fellow solders' wounds and needs. Not all memories and experiences from combat are painful, but all are lasting.

Our family's travels along this route of surprises continue, and it is our hope that many more generations will follow. That part of the story should be told, along with the stories of the many close friends and families we encountered along this journey.

Abbreviations

AAR – After Action Review
ACAV – Armored Cavalry Assault Vehicle
ADC – Assistant Division Commander
AFDC – Aid to Families with Dependent Children)
AOR – Area of Responsibility
AWAG – Americans Working Around the Globe
AWS – Amphibious Warfare School
BOQ – Bachelor Officer Quarters
CAS – Contemporary Affairs Seminar
CENTCOM – U.S. Central Command, responsible for Middle East Region
CG – Commanding General
CIB – Combat Infantryman's Badge
CONARC – Continental Army Command
DAS – Director of the Army Staff
DCSPER – Deputy Chief of Staff for Personnel
DCSOPS – Deputy Chief of Staff for Operations (the Pentagon)
DTIG – Deputy to the Inspector General
DoD – Department of Defense
DODSS – Department of Defense School System
EUCOM – European Command
FORSCOM – Forces Command

FRAGO – Fragmentary Order
GOMO – General Officer Management Office
HHC – Headquarters and Headquarters Company
IG – Inspector General
JAG – Judge Advocate General
JCS – Joint Chiefs of Staff
JROTC – Junior Reserve Officers Training Program
MACV – Military Assistance Command, Vietnam
MARS call – Military Auxiliary Radio System call
MRR – Motorized Rifle Regiment
NATO – North Atlantic Treaty Organization
NCAA – National Collegiate Athletic Association
NCO – noncommissioned officer
NTC – National Training Center
NVA – North Vietnamese Army
OCS – Officer Candidate School
OER – Officer Evaluation Report
OPFOR – opposing forces
PHV – Patrick Henry Village
POMCUS – Prepositioning of Materiel Configured in Unit Sets
PSD – Personal Security Detachment
PT – physical training
PX – Air Force Exchange System
ROTC – Reserve Officers Training Corps
R&R – Rest and Relaxation
SECDEF – Secretary of Defense
SES – Senior Executive Service
SFC – Sergeant First Class
TIG – The Inspector General
TJAG – The Judge Advocate General of the Army
TRADOC – Training and Doctrine Command